PERCEPTUAL METRICS FOR IMAGE DATABASE NAVIGATION

THE KLUWER INTERNATIONAL SERIES IN ENGINEERING AND COMPUTER SCIENCE

ROBOTICS: VISION, MANIPULATION AND SENSORS
Consulting Editor
Takeo Kanade

Other books in the series:

DARWIN2K: An Evolutionary Approach to Automated Design for Robotics
C. Leger
ISBN: 0-7923-7979-2
ENGINEERING APPROACHES TO MECHANICAL AND ROBOTIC DESIGN FOR MINIMALLY INVASIVE SURGERIES
A. Faraz, S. Payandeh
ISBN: 0-7923-7792-3
ROBOT FORCE CONTROL
B. Siciliano, L. Villani
ISBN: 0-7923-7733-8
DESIGN BY COMPOSITION FOR RAPID PROTOTYPING
M. Binnard
ISBN: 0-7923-8657-4
TETROBOT: A Modular Approach to Reconfigurable Parallel Robotics
G.J. Hamlin, A.C. Sanderson
ISBN: 0-7923-8025-8
INTELLIGENT UNMANNED GROUND VEHICLES: Autonomous Navigation Research at Carnegie Mellon
M. Hebert, C. Thorpe, A. Stentz
ISBN: 0-7923-9833-5
INTERLEAVING PLANNING AND EXECUTION FOR AUTONOMOUS ROBOTS
Illah Reza Nourbakhsh
ISBN: 0-7923-9828-9
GENETIC LEARNING FOR ADAPTIVE IMAGE SEGMENTATION
Bir Bhanu, Sungkee Lee
ISBN: 0-7923-9491-7
SPACE-SCALE THEORY IN EARLY VISION
Tony Lindeberg
ISBN 0-7923-9418
NEURAL NETWORK PERCEPTION FOR MOBILE ROBOT GUIDANCE
Dean A. Pomerleau
ISBN: 0-7923-9373-2
DIRECTED SONAR SENSING FOR MOBILE ROBOT NAVIGATION
John J. Leonard, Hugh F. Durrant-Whyte
ISBN: 0-7923-9242-6
A GENERAL MODEL OF LEGGED LOCOMOTION ON NATURAL TERRAINE
David J.Manko
ISBN: 0-7923-9247-7
INTELLIGENT ROBOTIC SYSTEMS: THEORY, DESIGN AND APPLICATIONS
K. Valavanis, G. Saridis
ISBN: 0-7923-9250-7

PERCEPTUAL METRICS FOR IMAGE DATABASE NAVIGATION

by

Yossi Rubner
Stanford University

Carlo Tomasi
Stanford University

KLUWER ACADEMIC PUBLISHERS
Boston / Dordrecht / London

Distributors for North, Central and South America:
Kluwer Academic Publishers
101 Philip Drive
Assinippi Park
Norwell, Massachusetts 02061 USA
Telephone (781) 871-6600
Fax (781) 681-9045
E-Mail <kluwer@wkap.com>

Distributors for all other countries:
Kluwer Academic Publishers Group
Distribution Centre
Post Office Box 322
3300 AH Dordrecht, THE NETHERLANDS
Telephone 31 78 6392 392
Fax 31 78 6546 474
E-Mail <services@wkap.nl>

 Electronic Services <http://www.wkap.nl>

Library of Congress Cataloging-in-Publication Data

Rubner, Yossi, 1964-
  Perceptual metrics for image database navigation / by Yossi Rubner, Carlo Tomasi.
    p. cm.—(The Kluwer international series in engineering and computer science; SECS 594)
  Includes bibliographical references and index.
  ISBN 0-7923-7219-0
    1. Database management. 2. Image processing—Digital techniques. I. Tomasi, Carlo,
  1956- II. Title. III. Series.

QA76.9.D3 R838 2000
005.74—dc21
                                                                    00-048691

Contents

List of Figures

List of Tables

Preface

The increasing amount of information available in today's world raises the need to retrieve relevant data efficiently. Unlike text-based retrieval, where keywords are successfully used to index into documents, content-based image retrieval poses up front the fundamental questions how to extract useful image features and how to use them for intuitive retrieval. We present a novel approach to the problem of navigating through a collection of images for the purpose of image retrieval, which leads to a new paradigm for image database search. We summarize the appearance of images by distributions of color or texture features, and we define a metric between any two such distributions. This metric, which we call the "Earth Mover's Distance" (EMD), represents the least amount of work that is needed to rearrange the mass is one distribution in order to obtain the other. We show that the EMD matches perceptual dissimilarity better than other dissimilarity measures, and argue that it has many desirable properties for image retrieval. Using this metric, we employ Multi-Dimensional Scaling techniques to embed a group of images as points in a two- or three-dimensional Euclidean space so that their distances reflect image dissimilarities as well as possible. Such geometric embeddings exhibit the structure in the image set at hand, allowing the user to understand better the result of a database query and to refine the query in a perceptually intuitive way. By iterating this process, the user can quickly zoom in to the portion of the image space of interest. We also apply these techniques to other modalities such as mug-shot retrieval.

Acknowledgments

This book is based on my Ph.D. thesis completed at Stanford University. I would like to thank the many people who made my time at Stanford a period I will treasure.

First and foremost, I would like to thank my advisor, Professor Carlo Tomasi, for guiding me through the Ph.D. program. From him I learned how to think critically, how to select problems, how to solve them, and how to present their solutions. His drive for scientific excellence has pushed me to aspire for the same. I thank him also for the seemingly impossible task of teaching me how to write clearly. Aside from being my advisor, he was a colleague, a teacher, and most important – a friend.

I would like to thank Professor Leonidas Guibas whom I consider my secondary advisor. He was always there to provide words of advice and to discuss new research ideas. His insights always enlightened me.

I would like to thank the other members of my oral committee, Professor Brian Wandell, Professor Rajeev Montwani, and Professor Shmuel Peleg from Hebrew university in Israel who was also a member of my reading committee. Special thanks to Professor Peleg for his willingness to participate in my oral defense from Germany by video-conferencing. I would also like to thank Professor Walter Murray and Professor Serge Plotkin for inspiring discussions on numerical and computational aspects of the EMD.

I was fortunate to interact with many people at Stanford and at various other places. These interaction shaped in many ways my understanding of computer vision. I would like to thank all the members and visitors of the vision group with whom I interacted, especially Stan Birchfield, Anat Caspi, Scott Cohen, Aris Gionis, Joachim Hornegger, Roberto Manduchi, Chuck Richards, Brian Rogoff, Mark Ruzon, Donald Tanguay, John Zhang, and Song-Chun Zhu. Special thanks to Mark for being willing to proof-read this book. I would like also to thank Professor Yaakov (Toky) Hel-Or and Professor Hagit (Gigi) Hel-Or for introducing me to the computer vision community and helping me with my first steps in the field, and Professor Ron Kimmel for much good advice. Thanks for Joern Hoffmann for working with me on the derivation of the Gabor filters, and to Jan Puzicha from the university of Bonn for an interesting and fruitful collaboration.

During my studies, I had the opportunity to work at Xerox Palo Alto Research Center. I would like to thank Professor Guibas for bringing me there, Dr. David Marimont, whom I consider as my third advisor, for his endless great ideas and support, and Dr. Michael Black, Dr. Tom Breuel, Dr. Don Kimber, Dr. Ruth Rosenholtz, and Dr. Randy Trigg for many interesting discussions.

Last but not least, I would like to thank my family. I am grateful to my parents, Zahava and David, who instilled in me the basic interest in science. Their encouragement and support helped me reach the goal of finishing a Ph.D. I would like to thank my daughters, Keren and Shir. They brought new dimensions to my life and keep reminding me that there is much more to life than my work. Above all I am grateful to my wife, Anat, for her love, support, patience and encouragement. To her I dedicate this book.

I was supported by the DARPA grant for Stanford's image retrieval project (contract DAAH04-94-G-0284).

Yossi Rubner

Introduction

The last thing that we discover in writing a book is to know what to put at the beginning.
—Pascal, 1623–1662

1. BACKGROUND AND MOTIVATION

Recent technological advances in disparate fields of endeavor have combined to make large databases of digital images accessible. These advances include:

1 Image acquisition devices, such as scanners and digital cameras,

2 Storage units that provide larger capacities for lower costs,

3 Access to enormous numbers of images via the internet, and the rapid growth of the World-Wide Web where images can be easily added and accessed.

Rummaging through such a large collection of images in search of a particular picture is unrewarding and time-consuming. Image database retrieval research attempts to automate parts of the tedious task of retrieving images that are similar to a given description.

Occasionally, semantic keywords are attached to the images. This can be done either by manually annotating the images or by automatically extracting the keyword from the context, for instance, from the the images' captions. When available, such keywords greatly assist the search. Practically, often images lack keywords and only the appearance features of the images can be used. Appearance features are useful even when semantic keywords are available because the keywords usually don't describe all the information in the image. The caption of a picture of the president walking in a garden is unlikely to specify the kinds of flowers in the garden, which may happen to be what the user is looking for.

The question of image similarity[1], the core of any retrieval system, is complex and delicate. The preferred mode of querying an image database is semantic. For example, we might search for images of children playing in a park. To satisfy such a query, the system must be able to automatically recognize children and parks in the images. Unfortunately, even after almost 50 years of computer vision, this level of image interpretation is still out of the question, and we must make do with similarity of appearance. More specifically, similarity between images can be defined by image features such as color, texture, or shape, and on the composition of these features in an image.

The discrepancy between the semantic query that the user has in mind and the syntactic features used to describe it makes it hard both for the user to specify the query, and for the system to return the correct images. Until semantic image interpretation can be done automatically, image retrieval systems cannot be expected to find the correct images. Instead, they should strive for a significant reduction in the number of images that the user needs to consider, and provide tools to view these images quickly and efficiently. The number of images can be reduced by extracting perceptually meaningful features and using dissimilarity measures that agree with perceptual similarity. Displaying the resulting images in an intuitive way can assist the user to quickly assess the query result.

A query into an image database is often formulated by sketching the desired feature or by providing an example of a similar image. Yet often we do not know the precise appearance of the desired image(s). We may want a sunset, but we do not know if sunsets in the database are on beaches or against a city skyline. When looking for unknown images, browsing, not query, is the preferred search mode. And the key requirement for browsing is that similar images are located nearby. Current retrieval systems list output images in order of increasing distance from the query. However, the distances among the returned images also convey useful information for browsing.

2.   PREVIOUS WORK

The increasing availability of digital imagery has created the need for content-based image retrieval systems, while the development of computation and storage resources provides the means for implementing them. This led to extensive research that resulted, in less than a decade, into numerous commercial and research-based systems, including QBIC [62], Virage [3], Photobook [66], Excalibur [26], Chabot [23], and VisualSEEK [92]. These systems allow the user to formulate queries using combinations of low-level image features such as color, texture, and shape. The queries are specified explicitly by providing the

---

[1]While it is more natural to use the term *similarity* in the context of perception, the measures discussed in this book actually compute the amount of *dissimilarity* between images. These terms are interchangeable as it is trivial to convert between similarities and dissimilarities. We use both terms according to the context.

desired feature values or implicitly by specifying an example image. Some systems also use the spatial organization of the image features, so that similarity is determined not only by the existence of certain features, but also by their absolute or relative location in the image [40, 3, 95, 92, 18, 12]. Early systems focused on the search engine (given a query, find the best matches) and did not use previous queries to understand better what the user was looking for. Recent systems allow the user to refine the search by indicating the relevance (or irrelevance) of images in the returned set. This is known as *relevance feedback* [81, 59, 92].

In [41], the authors provide an in-depth review of content-based image retrieval systems (CBIR). They also identify a number of unanswered key research questions, including the development of more robust and compact image content features and dissimilarity measures that model perceptual similarity more accurately. We approach these issues in Chapters 1-5 and the problem of expanding a content-based image search engine to an intuitive navigation system in Chapters 6 and 7.

3.   OVERVIEW OF THE BOOK

We present a novel framework for computing the distance between images and a set of tools to visualize parts of the database and browse its content intuitively. In particular, we address the following questions:

- What features describe the content of an image well?

- How to summarize the distribution of these features over an image?

- How to measure the dissimilarity between distributions of features?

- How can we effectively display the results of a search?

- How can a user browse the images in the database in an intuitive and efficient way?

In this book, we focus on the overall color and texture content of images as the main criterion for similarity. The overall distribution of colors within an image contributes to the mood of the image in an important way and is a useful clue for the content of an image. Sunny mountain landscapes, sunsets, cities, faces, jungles, candy, and fire fighter scenes lead to images that have different but characteristic color distributions. While color is a property of single pixels, texture describes the appearance of bigger regions in the image. Often, semantically similar objects can be characterized by similar textures.

Summarizing feature distributions has to do with perceptual significance, invariance, and efficiency. Features should be represented in a way that reflects a human's appreciation of similarities and differences. At the same time, the distributions of the image features should be represented by a collection of data that is small, for efficiency, but rich enough to reproduce the essential information. The issue of relevance to human perception has been resolved by the choice of appropriate representations. For color we choose the CIELab color space, and for texture we use Gabor filters. For both we discuss their relations with perceptual similarity. We summarize the feature distributions by a small collection of weighted points in feature space where the number of points adapts to capture the complexity of the distributions; we call this set of points a *signature*.

Defining a dissimilarity measure between two signatures first requires a notion of distance between the basic features that are aggregated into the signatures. We call this distance the *ground distance*. For instance, in the case of color, the ground distance measures dissimilarity between individual colors. We address the problem of lifting these distances from individual features to full distributions. In other words, we want to define a consistent measure of distance, or dissimilarity, between two distributions in a space endowed with a ground distance. We introduce the *Earth Mover's Distance* as a useful and flexible metric, thereby addressing the question of image similarity.

If the pictures in a database can be spatially arranged so that their locations reflect their differences and similarities, browsing the database by navigating in this space becomes intuitively meaningful. In fact, the database is now endowed with a metric structure, and can be explored with a sense of continuity and comprehensiveness. Parts of the database that have undesired distributions need not be traversed; on the other hand, interesting regions can be explored with a sense of getting closer or farther away from the desired distribution of colors. In summary, the user can form a mental, low-detail picture of the entire database, and a more detailed picture of the more interesting parts of it.

4.   ROAD MAP

The rest of the book is organized as follows: Chapter 1 addresses the issue of summarizing distributions of features and surveys some of the most commonly used distribution-based dissimilarity measures. In Chapter 2 we introduce the Earth Mover's Distance, which we claim is a preferred dissimilarity measure for image retrieval, and discuss its properties. In Chapters 3 and 4, respectively, we define color and texture features and show that, combined with the EMD, they lead to effective image retrieval. In Chapter 5 we conduct extensive experiments where we compare the retrieval results, for color and texture, using various dissimilarity measures. The problem of displaying the results of a search in a useful way is approached in Chapter 6. Finally, in Chapter 7 we extend our display technique to a full navigation system that allows intuitive refinement of a query.

5.   THE ATTACHED CD

Due to technical limitations, this book was printed in black-and-white. Since color is one of the more important visual features used in this work, we attach a CD that contains a color version of the book in PostScript format. Figures in the text that are originally in color, have a mention to this in their caption. Please refer to the CD for the color versions of the figures.

Chapter 1

DISTRIBUTION-BASED DISSIMILARITY MEASURES

The difference between the right word and a similar word is the difference between light-ning and a lightning bug.

—Mark Twain, 1835–1910

In order for an image retrieval system to find images that are visually similar to the given query, it should have both a proper representation of the images visual features and a measure that can determine how similar or dissimilar the different images are from the query. Assuming that no textual captions or other manual annotations of the images are given, the features that can be used are descriptions of the image content, such as color [97, 62, 96, 91, 4], texture [25, 62, 6, 69, 56, 4], and shape [42, 62, 31, 44]. These features usually vary substantially over an image, both because of inherent variations in surface appearance and as a result of changes in illumination, shading, shadowing, foreshortening, etc. Thus, the appearance of a region is better described by the *distribution* of features, rather than by individual feature vectors.

Dissimilarity measures, based on empirical estimates of the distributions of features, have been developed and used for different tasks in low-level computer vision including classification [63], image retrieval [28, 97, 71, 80] and segmentation [32, 39].

In this chapter, we first describe different representations for distributions of image features and discuss their advantages and disadvantages. We then survey and categorize some of the most commonly used dissimilarity measures.

1.   REPRESENTING DISTRIBUTIONS OF FEATURES

1.1.   Histograms

A *histogram* $\{h_\mathbf{i}\}$ is a mapping from a set of d-dimensional integer vectors i to the set of non-negative reals. These vectors typically represent bins (or their centers) in a fixed partitioning of the relevant region of the underlying feature space. The associated reals are a measure of the mass of the distribution that falls into the corresponding bin. For instance, in a grey-level histogram, d is equal to one, the set of possible grey values is split into N intervals, and $h_\mathbf{i}$ is the number of pixels in an image that have a grey value in the interval indexed by i (a scalar in this case).

The fixed partitioning of the feature space can be *regular*, with all bins having the same size. A major problem of regular histograms is that the number of bins grows exponentially with the number of dimensions, affecting storage and computational costs. If the distribution of features of all the images is known *a priori*, then *adaptive binning* can be used, whereby the location and size of the histogram bins are adapted to the distribution. The binning is induced by a set of *prototypes* $\{c_\mathbf{i}\}$ and the corresponding Voronoi tessellation. Adaptive histograms are formally defined by

$$h_i = \left| \{ \mathbf{x} \ : \ i = \arg\min_j \| I(\mathbf{x}) - c_j \| \} \right| .$$

Here $I(\mathbf{x})$ denotes the feature vector at image position $\mathbf{x}$, and $|\cdot|$ is the number of elements in a set. The histogram entry h_i corresponds to the number of image pixels in bin i. Adaptive histograms usually have one-dimensional index as the ordering of the bins in space is not well defined. A suitable set of prototypes can be determined by a vector quantization procedure, *e.g.* K-means (see [61] for a review).

For images that contain a small amount of information, a finely quantized histogram is highly inefficient. On the other hand, for images that contain a large amount of information, a coarsely quantized histogram would be inadequate. Because histograms are fixed-size structures, they cannot achieve a good balance between expressiveness and efficiency.

1.2.   Signatures

Unlike histograms, whose bins are defined over the entire database, the clusters in *signatures* are defined for each image individually. A signature $\{s_j = (m_j, w_{m_j})\}$ represents a set of feature clusters. Each cluster is represented by its mean (or mode) m_j and the fraction w_{m_j} of pixels that belong to that cluster. The integer subscript j ranges from one to a value that varies with the complexity of the particular image. While j is simply an integer, the representative m_j is a d-dimensional vector. In general, the same vector quantization algorithms that are used to compute adaptive histograms can be used for clustering, as long as they are applied to every image independently, adapting the number of clusters to the complexity of the individual images. Simple images have short signatures while complex images have long ones. An example of a color signature is given in Figure 1.1. Part (a) shows a color image, and part (b) shows the corresponding clusters in color space. The cluster weights are reflected in the sizes of the spheres. This is a colorful image so its signature is large. More examples of color and texture signatures can be found in Chapters 3 and 4.

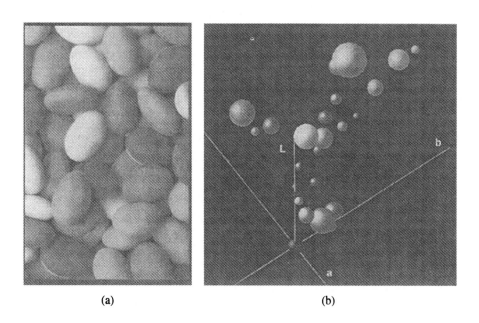

(a) (b)

Figure 1.1. (a) Color image. (b) Its color signature. *(This is a color figure)*

A histogram $\{h_i\}$ can be viewed as a signature $\{s_j = (m_j, w_{m_j})\}$ in which the vectors i index a set of clusters defined by a fixed *a priori* partitioning of the underlying space. If vector i maps to cluster j, the point m_j is the central value in bin i of the histogram, and w_j is equal to h_i.

We show in Chapters 3 and 4 that representing the content of an image database with signatures leads to better query results than with histograms. This is the case even when the signatures contain, on the average, significantly less information than the histograms. By "information" here we refer to the minimal number of bits needed to code the signatures and the histograms.

1.3. Other representations

In addition to histograms and signatures, which are based on global or local tessellation of the space into non-overlapping regions, there are other techniques to describe non-parametric distributions. For example, in kernel density estimation [22], each data point is replaced by some kernel (Parzen window) and the density estimations is regarded as the superposition of all these kernels. Representing and comparing kernel density estimation becomes unwieldy and computationally expensive as the amount of data and the dimensionality of the feature space becomes large.

2. HISTOGRAM-BASED DISSIMILARITY MEASURES

Most of the dissimilarity measures used in image retrieval measure the dissimilarity between two histograms $H = \{h_i\}$ and $K = \{k_i\}$. We divide these measures into two categories. The *bin-by-bin* dissimilarity measures compare contents of corresponding histogram bins, that is, they compare h_i and k_i for all i, but not h_i and k_j for $i \neq j$. The *cross-bin* measures also compare non-corresponding bins. Cross-bin distances make use of the *ground distance* d_{ij}, defined as the distance between the representative features for bin i and bin j. Predictably, bin-by-bin measures are more sensitive to the position of bin boundaries. We start with a few definitions.

2.1. Definitions

Here we define terms that are used with regard to the dissimilarity measures defined is this chapter.

2.1.1 Metric Space

A space $\mathcal{A}$ is called a metric space if for any of its two elements x and y, there is a number $\rho(x, y)$, called the distance, that satisfies the following properties

- $\rho(x, y) \geq 0$　(non-negativity)

- $\rho(x, y) = 0$ if and only if $x = y$　(identity)

- $\rho(x, y) = \rho(y, x)$　(symmetry)

- $\rho(x, z) \leq \rho(x, y) + \rho(y, z)$　(triangle inequality)

2.1.2 Partial Matching

With partial matching, when the total mass of one distribution is smaller than that of the other, the dissimilarity score is computed only with respect to the most similar part of the larger distribution. This is a useful property for image retrieval since often only part of the image or the features are known or are of interest. Partial matches also provide a way to deal with occlusions and clutter in image retrieval.

2.2. Bin-by-bin dissimilarity measures

In this category only pairs of bins in the two histograms that have the same index are matched. The dissimilarity between two histograms is a combination of all the pairwise comparisons. A ground distance is used by these measures only implicitly and in an extreme form: features that fall into the same bin are close enough to each other to be considered the same, and those that do not are too far apart to be considered similar. In this sense, bin-by-bin measures imply a binary ground distance with a threshold depending on bin size.

2.2.1 Minkowski-form distance

The Minkowski-form distance is defined based on the L_p norm as

$$d_{L_p}(H, K) = \left(\sum_i |h_i - k_i|^p \right)^{1/p} .$$

The L_1 distance is often used for computing dissimilarity between color images [97]. Other common usages are L_2 and L_∞ (*e.g.* for texture dissimilarity [102]). In [96] it was shown that for image retrieval the L_1 distance results in many false negatives because neighboring bins are not considered.

2.2.2 Histogram intersection

Histogram intersection [97] is defined by

$$d_\cap(H, K) = 1 - \frac{\sum_i \min(h_i, k_i)}{\sum_i k_i} .$$

It is attractive because of its ability to handle partial matches when the area of one histogram (the sum over all the bins) is smaller than that of the other. It is shown in [97] that when the areas of the two histograms are equal, the histogram intersection is equivalent to the (normalized) L_1 distance.

2.2.3 Kullback-Leibler divergence and Jeffrey divergence

The Kullback-Leibler (KL) divergence [50] is defined as:

$$d_{KL}(H, K) = \sum_i h_i \log \frac{h_i}{k_i} .$$

From an information theoretic point of view, the K-L divergence measures how inefficient on average it would be to code one histogram using the other as the code-book [13]. However, the K-L divergence is non-symmetric and is sensitive to histogram binning. The empirically derived Jeffrey divergence is a modification of the K-L divergence that is numerically stable, symmetric and robust with respect to noise and the size of histogram bins [71]. It is defined as:

$$d_J(H, K) = \sum_i \left(h_i \log \frac{h_i}{m_i} + k_i \log \frac{k_i}{m_i} \right) ,$$

where $m_i = \frac{h_i+k_i}{2}$.

2.2.4 χ^2 statistics

The χ^2 statistics is defined as

$$d_{\chi^2}(H, K) = \sum_i \frac{(h_i - m_i)^2}{m_i} ,$$

where again $m_i = \frac{h_i+k_i}{2}$. This quantity measures how unlikely it is that one distribution was drawn from the population represented by the other.

2.2.5   Drawbacks of bin-by-bin dissimilarity measures

These dissimilarity definitions are appropriate in different areas. For example, the Kullback-Leibler divergence is justified by information theory and the χ^2 statistics by statistics. However, these measures do not necessarily match perceptual similarity well. Their major drawback is that they account only for the correspondence between bins with the same index, and do not use information across bins. This problem is illustrated in Figure 1.2(a) which shows two pairs of one-dimensional gray scale histograms. Although the two histograms on the left are the same except for a shift by one bin, the L_1 distance (as a representative of bin-by-bin dissimilarity measures) between them is larger than the L_1 distance between the two histograms on the right, in contrast to perceptual dissimilarity. This can be fixed by using correspondences between bins in the two histograms and the ground distance between them as shown in part (b) of the figure.

Another drawback of bin-by-bin dissimilarity measures is their sensitivity to bin size. A binning that is too coarse will not have sufficient discriminative power, while a binning that is too fine might place similar features in different bins that will not be matched. On the other hand, cross-bin dissimilarity measures, described next, always yield better results when the bins get smaller.

2.3.   Cross-bin dissimilarity measures

When a ground distance that matches perceptual dissimilarity is available for single features, incorporating this information results in perceptually more meaningful dissimilarity measures for distributions of features.

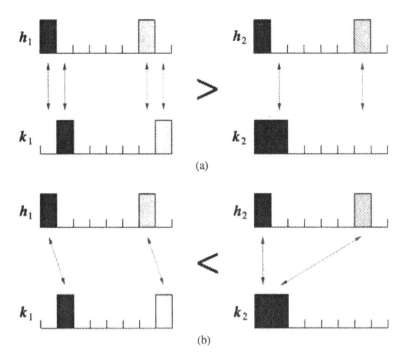

(a)

(b)

Figure 1.2. An example where the L_1 distance does not match perceptual dissimilarity. (a) Assuming that histograms have unit mass $d_{L_1}(\mathbf{h}_1, \mathbf{k}_1) = 2$, $d_{L_1}(\mathbf{h}_2, \mathbf{k}_2) = 1$. (b) Perceptual dissimilarity is based on correspondence between bins in the two histograms.

2.3.1 Quadratic-form distance

$$d_A(H, K) = \sqrt{(\mathbf{h} - \mathbf{k})^T \mathbf{A}(\mathbf{h} - \mathbf{k})} \,,$$

where $\mathbf{h}$ and $\mathbf{k}$ are vectors that list all the entries in H and K. This distance was suggested for color-based retrieval in [62].

Cross-bin information is incorporated via a similarity matrix $\mathbf{A} = [a_{ij}]$ where a_{ij} denotes similarity between bins i and j. Here i and j are sequential (scalar) indices into the bins.

For our experiments, we followed the recommendation in [62] and used $a_{ij} = 1 - d_{ij}/d_{max}$ where d_{ij} is the ground distance between the feature descriptors of bins i and j of the histogram, and $d_{max} = \max_{ij} d_{ij}$. Although in general the quadratic-form distance is not a metric, it can be shown that with this choice of $\mathbf{A}$ it is indeed a metric.

The quadratic-form distance does not enforce a one-to-one correspondence between mass elements in the two histograms: The same mass in a given bin of the first histogram is simultaneously made to correspond to masses contained in different bins of the other histogram. This is illustrated in Figure 1.3(a),

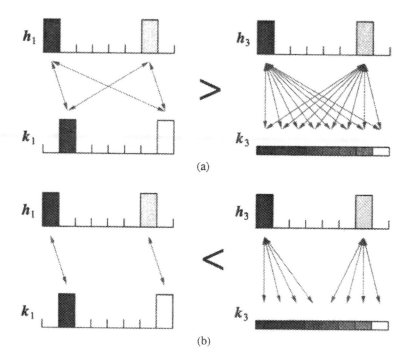

Figure 1.3. An examples where the quadratic-form distance does not match perceptual dissimilarity. (a) Assuming that histograms have unit mass $d_A(\mathbf{h_1}, \mathbf{k_1}) = 0.1429$, $d_A(\mathbf{h_3}, \mathbf{k_3}) = 0.0893$. (b) Perceptual dissimilarity is based on correspondence between bins in the two histograms

where the quadratic-form distance between the two histograms on the left is larger than the distance between the two histograms on the right. Again, this is clearly at odds with perceptual dissimilarity. The desired distance here should be based on the correspondences shown in part (b) of the figure.

Similar conclusions were obtained in [96], where it was shown that using the quadratic-form distance in image retrieval results in false positives, because it tends to overestimate the mutual similarity of color distributions without a pronounced mode.

2.3.2   One-dimensional match distance

$$d_M(H, K) = \sum_i |\hat{h}_i - \hat{k}_i| , \qquad (1.1)$$

where $\hat{h}_i = \sum_{j \leq i} h_j$ is the cumulative histogram of $\{h_i\}$, and similarly for $\{k_i\}$.

The match distance [87, 105] between two one-dimensional histograms is defined as the L_1 distance between their corresponding cumulative histograms. For one-dimensional histograms with equal areas, this distance is a special case of the EMD, which we present in Chapter 2, with the important differences that the match distance cannot handle partial matches or other ground distances. The one-dimensional match distance does not extend to higher dimensions because the relation $j \leq i$ is not a total ordering in more than one dimension, and the resulting arbitrariness causes problems. The match distance is extended in [105] for multi-dimensional histograms by using graph matching algorithms. This extension is similar in spirit to the EMD, which can be seen as a generalization of the match distance. In the rest of this chapter, the term "match distance" refers only to the one-dimensional case, for which analytic formulation is available.

2.3.3 Kolmogorov-Smirnov statistics

$$d_{KS}(H, K) = \max_i(|\hat{h}_i - \hat{k}_i|) .$$

Again, $\hat{h}_i$ and $\hat{k}_i$ are cumulative histograms.

The Kolmogorov-Smirnov statistics is a common statistical measure for un-binned distributions. Although, for consistency, we use here cumulative histograms, the Kolmogorov-Smirnov statistics is defined on the cumulative distributions so that no binning is required and it can be applied to continues data. In the case of the null hypothesis (data sets drawn from the same distribution) the distribution of the Kolmogorov-Smirnov statistics can be calculated, thus giving the significance of the result [16]. Similarly to the match distance, it is defined only for one dimension.

2.4. Parameter-based dissimilarity measures

These methods first compute a small set of parameters from the histograms, either explicitly or implicitly, and then compare these parameters. For instance, the distance between distributions is computed in [96] as the sum of the weighted distances of the distributions' first three moments. In [18], only the peaks of color histograms are used for color image retrieval. In [53], textures are compared based on measures of their periodicity, directionality, and randomness, while in [56] texture distances are defined by comparing their means and standard deviations in a weighted-L_1 sense.

Additional dissimilarity measures for image retrieval are evaluated and compared in [92, 71].

	L_r	HI	KS	KL	JD	χ^2	QF	MD	EMD
Symmetric	+	-	+	-	+	+	+	+	+
Triangle inequality	+	-	+	-	-	-	(1)	+	(1)
Exploits ground distance	-	-	+	-	-	-	+	+	+
Signatures	-	-	-	-	-	-	-	-	+
Multiple dimensions	+	+	-	+	+	+	+	-	+
Partial matches	-	+	-	-	-	-	-	-	+
Computational complexity	low	low	low	low	low	low	high	low	high

Table 1.1. Properties of the different dissimilarity measures: Minkowski-form (L_r), histogram intersection (HI), Kolmogorov-Smirnov (KS), Kullback-Leibler (KL), Jeffrey divergence (JD), χ^2 statistics (χ^2), quadratic form (QF), match distance (MD), and the Earth Mover's Distance (EMD). (1) The triangle inequality only holds for specific ground distances.

2.5.   Properties

Table 1.1 compares the main properties of the different measures presented in Sections 2.3 and 2.2. For comparison, we include the EMD, which is defined in Chapter 2.

The Kolmogorov-Smirnov statistics and the match distance are defined only for one-dimensional distributions. Thus, they cannot be used for color and texture.

Metric dissimilarity measures enable more efficient indexing algorithms for image databases, as the triangle inequality entails lower bounds that can be exploited to substantially alleviate the computational burden. The triangle inequality does not hold for the histogram intersection, Kolmogorov-Smirnov, Jeffrey divergence, and χ^2 statistics. The quadratic form is metric only for some ground distances (see [35]), while the EMD is metric only when the ground distance is metric and if the two signatures have the same total weight. All the evaluated measures are symmetric except for the histogram intersection and the Kullback-Leibler divergence.

A useful property for image retrieval is the ability to handle *partial matches*. Only the histogram intersection and the EMD directly allow partial matches.

Computational complexity is another important consideration. The computational complexity of the quadratic form and the EMD are the highest among the evaluated measures. There are good approximations for the quadratic form that can be computed efficiently [62]. Computing the EMD for large histograms is infeasible for an online image retrieval system, as for every dissimilarity calculation a linear optimization is necessary. However, as we show in Chapters 3 and 4, by using signatures instead of histograms, good results are achieved by the EMD even with small signatures and, consequently, less computation.

3. SUMMARY

In this chapter we presented methods for the representation of distributions of image features, with histograms being the most common method for image retrieval. We also surveyed and compared histogram-based dissimilarity measures.

Histograms are inflexible: they cannot achieve a good balance between expressiveness and efficiency. Signatures, on the other hand, adjust to the specific images. Unfortunately, most dissimilarity measures cannot be applied to signatures. In the next chapter we present the Earth Mover's Distance, which is designed for signatures, and show that it has many desirable properties for image retrieval.

Chapter 2

THE EARTH MOVER'S DISTANCE

The great God endows his children variously. To some He gives intellect– and they move the earth ...

—Mary Roberts Rinehart, 1876–1958

A ground distance between single visual image features can often be found by psychophysical experiments. For example, perceptual color spaces have been devised in which the Euclidean distance between two colors approximately matches the human perception of their difference. Measuring perceptual distance becomes more complicated when sets of features, rather than single features, are being compared. In Section 2 we showed the problems caused by dissimilarity measures that do not handle correspondences between different bins in the two histograms. Such correspondences are key to a perceptually natural definition of the distances between sets of features. This observation led to distance measures based on bipartite graph matching [65, 108], defined as the minimum cost of matching elements between the two histograms.

In [65] the distance between two gray scale images is computed as follows: every pixel is represented by n "pebbles" where n is an integer representing the gray level of that pixel. After normalizing the two images to have the same number of pebbles, the distance between them is computed as the minimum cost of matching the pebbles between the two images. The cost of matching two single pebbles is based on their distance in the image plane. In this section we take a similar approach and derive the *Earth Mover's Distance*[1] (EMD) as a useful metric between signatures for image retrieval in different feature spaces.

---

[1]The name EMD was suggested by Stolfi [94].

The main difference between the two approaches is that we solve the transportation problem which finds the optimal match between two distributions where variable-sized pieces of "mass" are allowed to be moved together, in contrast to the matching problem where unit elements of fixed size are matched individually. This distinction significantly increases efficiency, due to the more compact representation as a result of clustering pixels in the feature space. The implementation is fast enough for online image retrieval systems. In addition, as we will show, our formulation allows for partial matches, which are important for image retrieval applications. Finally, instead of computing image distances based on the cost of moving pixels in the image space, where the ground distance is perceptually meaningless, we compute distances in feature spaces, where the ground distances can be perceptually better defined.

Intuitively, given two distributions, one can be seen as piles of earth in feature space, the other as a collection of holes in that same space. The EMD measures the least amount of work needed to fill the holes with earth. Here, a unit of work corresponds to transporting a unit of earth by a unit of ground distance.

Computing the EMD is based on a solution to the well-known *transportation problem* [38], also known as the Monge-Kantorovich mass transference problem, which goes back to 1781 when it was first introduced by Monge [60] in the following way:

> Split two equally large volumes into infinitely small particles and then associate them with each other so that the sum of products of these paths of the particles to a volume is least. Along what paths must the particles be transported and what is the smallest transportation cost?

See [72] for an excellent survey of the history of the Monge-Kantorovich mass transference problem. This distance was first introduced to the computer vision community by [105].

Suppose that several *suppliers*, each with a known amount of goods, are required to supply several *consumers*, each with a known capacity. For each supplier-consumer pair, the cost of transporting a single unit of goods is given. The transportation problem is, then, to find a least-expensive flow of goods from the suppliers to the consumers that satisfies the consumers' demand. Signature matching can be naturally cast as a transportation problem by defining one signature as the supplier and the other as the consumer, and by setting the cost for a supplier-consumer pair to equal the ground distance between an element in the first signature and an element in the second. Intuitively, the solution is the minimum amount of "work" required to transform one signature into the other. In the following we will refer to the supplies as "mass."

In Section 1 we formally define the EMD. We discuss its properties in Section 2 and specify computational issues, including lower bounds, in Section 3. In Section 4 we describe a saturated ground distance and claim better agreement with psychophysics than simpler distance measures. Finally, in Section 5 we mention some extensions to the EMD.

1. DEFINITION

The EMD is based on the following linear programming problem: Let $P = \{(\mathbf{p}_1, w_{\mathbf{p}_1}), \ldots, (\mathbf{p}_m, w_{\mathbf{p}_m})\}$ be the first signature with m clusters, where $\mathbf{p}_i$ is the cluster representative and $w_{\mathbf{p}_i}$ is the weight of the cluster; $Q = \{(\mathbf{q}_1, w_{\mathbf{q}_1}), \ldots, (\mathbf{q}_n, w_{\mathbf{q}_n})\}$ the second signature with n clusters; and $\mathbf{D} = [d_{ij}]$ the ground distance matrix where $d_{ij} = d(\mathbf{p}_i, \mathbf{q}_j)$ is the ground distance between clusters $\mathbf{p}_i$ and $\mathbf{q}_j$.

We want to find a flow $\mathbf{F} = [f_{ij}]$, with f_{ij} the flow between $\mathbf{p}_i$ and $\mathbf{q}_j$, that minimizes the overall cost

$$\text{WORK}(P, Q, \mathbf{F}) = \sum_{i=1}^{m} \sum_{j=1}^{n} d(\mathbf{p}_i, \mathbf{q}_j) f_{ij} \, ,$$

subject to the following constraints:

$$f_{ij} \geq 0, \quad 1 \leq i \leq m, \quad 1 \leq j \leq n, \tag{2.1a}$$

$$\sum_{j=1}^{n} f_{ij} \leq w_{\mathbf{p}_i}, \quad 1 \leq i \leq m, \tag{2.1b}$$

$$\sum_{i=1}^{m} f_{ij} \leq w_{\mathbf{q}_j}, \quad 1 \leq j \leq n, \tag{2.1c}$$

$$\sum_{i=1}^{m} \sum_{j=1}^{n} f_{ij} = \min\left(\sum_{i=1}^{m} w_{\mathbf{p}_i}, \sum_{j=1}^{n} w_{\mathbf{q}_j}\right). \tag{2.1d}$$

Constraint (2.1a) allows moving "supplies" from P to Q and not vice versa. Constraint (2.1b) limits the amount of supplies that can be sent by the clusters in P to their weights. Constraint (2.1c) limits the clusters in Q to receive no more supplies than their weights; and Constraint (2.1d) forces the maximum amount of supplies possible to be moved. We call this amount the *total flow*. Once the transportation problem is solved, and we have found the optimal flow $\mathbf{F}$, the Earth Mover's Distance is defined as the resulting work normalized by the total flow:

$$\text{EMD}(P, Q) = \frac{\sum_{i=1}^{m} \sum_{j=1}^{n} d(\mathbf{p}_i, \mathbf{q}_j) f_{ij}}{\sum_{i=1}^{m} \sum_{j=1}^{n} f_{ij}} \, ,$$

The normalization factor is the total weight of the smaller signature, because of Constraint (2.1d). This factor is needed when the two signatures have different total weights, in order to avoid favoring smaller signatures. An example of a two-dimensional EMD is shown in Figure 2.1. In general, the ground distance can be any distance and will be chosen according to the problem at hand. Further discussion of the ground distance is given in Section 4.

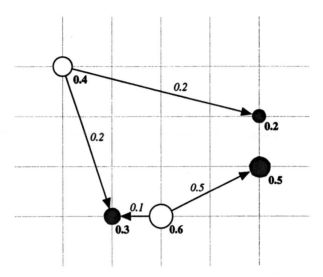

Figure 2.1. The Earth Mover's Distance in two dimensions between a signature with three points (black discs) and a signature with two points (white discs). The bold numbers attached to the discs are the weights of the points and the italic numbers on the arrows are the weights moved between the points. The sides of the grid cells are one unit long, and the resulting EMD is 2.675.

2. PROPERTIES

The EMD naturally extends the notion of a distance between single elements to that of a distance between sets, or distributions, of elements. The advantages of the EMD over previous distribution distance measures should now be apparent:

1 The EMD applies to signatures, which subsume histograms as shown in Section 1. The greater compactness and flexibility of signatures is in itself an advantage, and having a distance measure that can handle these variable-size structures is important.

2 The cost of moving "earth" reflects the notion of nearness properly, without the quantization problems of most current measures. Even for histograms, in fact, items from neighboring bins now contribute similar costs, as appropriate.

3 The EMD allows for partial matches in a natural way. This is important, for instance, in order to deal with occlusions and clutter in image retrieval applications, and when matching only parts of an image.

4 If the ground distance is a metric and the total weights of two signatures are equal, the EMD is a true metric, which endows image spaces with a metric

structure. Metric dissimilarity measures allow for efficient data structures and search algorithms [8, 10].

We now prove the final statement.

THEOREM 2.1 *If two signatures, P and Q, have equal weights and the ground distance* $d(\mathbf{p}_i, \mathbf{q}_j)$ *is metric for all* $\mathbf{p}_i$ *in P and* $\mathbf{q}_j$ *in Q, then EMD(P, Q) is also metric.*

Proof: To prove that a distance measure is metric, we must prove the following: positive definiteness (EMD$(P, Q) \geq 0$ and EMD$(P, Q) = 0$ iff $P \equiv Q$), symmetry (EMD$(P, Q) =$ EMD(Q, P)), and the triangle inequality (for any signature R, EMD$(P, Q) \leq$ EMD$(P, R) +$ EMD(R, Q)).

Positive definiteness and symmetry hold trivially in all cases, so we only need to prove that the triangle inequality holds. Without loss of generality we assume here that the total sum of the flows is 1. Let $\{f_{ij}\}$ be the optimal flow from P to R and $\{g_{jk}\}$ be the optimal flow from R to Q. Consider the flow $P \mapsto R \mapsto Q$. We now show how to construct a feasible flow from P to Q that represents no more work than that of moving mass optimally from P to Q through R. Since the EMD is the least possible amount of feasible work, this construction proves the triangle inequality.

The largest weight that moves as one unit from $\mathbf{p}_i$ to $\mathbf{r}_j$ and from $\mathbf{r}_j$ to $\mathbf{q}_k$ defines a flow which we call b_{ijk} where i, j and k correspond to $\mathbf{p}_i$, $\mathbf{r}_j$ and $\mathbf{q}_k$ respectively. Clearly $\sum_k b_{ijk} = f_{ij}$, the flow from P to R, and $\sum_i b_{ijk} = g_{jk}$, the flow from R to Q. We define

$$h_{ik} \stackrel{\triangle}{=} \sum_j b_{ijk}$$

to be a flow from $\mathbf{p}_i$ to $\mathbf{q}_k$. This flow is a feasible one because it satisfies the constraints (2.1a)-(2.1d) in Section 1. Constraint (2.1a) holds, since $b_{ijk} > 0$ by construction. Constraints (2.1b) and (2.1c) hold because

$$\sum_k h_{ik} = \sum_{j,k} b_{ijk} = \sum_j f_{ij} = w_{\mathbf{p}_i},$$

and

$$\sum_i h_{ik} = \sum_{i,j} b_{ijk} = \sum_j g_{jk} = w_{\mathbf{q}_k},$$

and constraint (2.1d) holds because the signatures have equal weights. Since EMD(P, Q) is the minimal flow from P to Q, and h_{ik} is some legal flow from

P to Q,

$$
\begin{aligned}
\text{EMD}(P, Q) &\leq \sum_{i,k} h_{ik} d(\mathbf{p}_i, \mathbf{q}_k) \\
&= \sum_{i,j,k} b_{ijk} d(\mathbf{p}_i, \mathbf{q}_k) \\
&\leq \sum_{i,j,k} b_{ijk} d(\mathbf{p}_i, \mathbf{r}_j) + \sum_{i,j,k} b_{ijk} d(\mathbf{r}_j, \mathbf{q}_k) \qquad (d(\cdot, \cdot) \text{ is metric}) \\
&= \sum_{i,j} f_{ij} d(\mathbf{p}_i, \mathbf{r}_j) + \sum_{j,k} g_{jk} d(\mathbf{r}_j, \mathbf{q}_k) \\
&= \text{EMD}(P, R) + \text{EMD}(R, Q) \ .
\end{aligned}
$$

∎

3.   COMPUTATION

It is important that the EMD be computed efficiently, especially if it is used for image retrieval systems where a quick response is essential. Fortunately, efficient algorithms for the transportation problem are available. We used the transportation-simplex method [37], a streamlined simplex algorithm [17] that exploits the special structure of the transportation problem. We would like signature P (with m clusters) and signature Q (with n clusters) to have equal weights. If the original weights are not equal, we add a *slack cluster* to the smaller signature and set to zero the associated costs of moving its mass. This gives the appropriate "don't care" behavior. It is easy to see that for signatures with equal weights, the inequality signs in Constraints 2.1b and 2.1c can be replaced by equality signs. This makes Constraint 2.1d redundant. The resulting linear program has now mn variables (the flow components f_{ij}) and $m + n$ constraints. With the regular simplex method this would result in a *tableau* with mn rows and $m + n$ columns. Exploiting the special structure of the transportation problem, the tableau can be compacted into the following table with only $m + 1$ rows and $n + 1$ columns:

$$
\begin{array}{|ccc|c|}
\hline
f_{11} & \cdots & f_{1n} & w_{\mathbf{p}_1} \\
\vdots & & \vdots & \vdots \\
f_{m1} & \cdots & f_{mn} & w_{\mathbf{p}_m} \\
\hline
w_{\mathbf{q}_1} & \cdots & w_{\mathbf{q}_n} & \\
\hline
\end{array}
$$

This table is much more efficient to create and maintain than the original tableau, which never has to be computed. In addition, the computation needed in the simplex optimization iterations is significantly reduced with the use of the transportation table.

A property of the transportation problem is that a feasible solution always exists and is easy to compute. A good initial basic feasible solution can drastically decrease the number of iterations needed. We compute the initial basic feasible solution by Russell's method [82].

The computational complexity of the transportation-simplex is based on the simplex algorithm, which has an exponential worst case [46]. However, the performance is much better in practice, because of the special structure in our case and the good initial solution. We empirically measure the time-performance of our EMD implementation by generating random signatures that range in size from 1 to 500. For each size we generate 100 pairs of random signatures and record the average CPU time for computing the EMD between them. The results are shown in Figure 2.2. This experiment was done on a SGI Indigo 2 with a 195MHz CPU.

Other methods that efficiently solve the transportation problem include interior-point algorithms [45], which have polynomial time complexity. By formalizing the transportation problem as the uncapacitated minimum cost network flow problem [2], it can be solved in our bipartite graph case in $O(n^3 \log n)$, where n is the number of clusters in the signatures[2]. This is consistent with the empirical running times of our algorithm, as can be inferred from Figure 2.2.

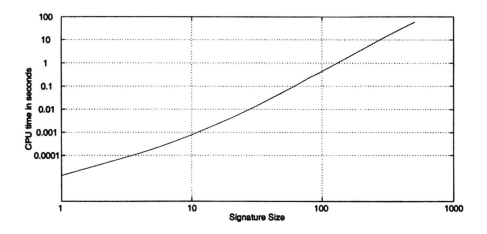

Figure 2.2. A log-log plot of the average computation time for random signatures as a function of signature size.

---

[2]This bound assumes that the two signatures have the same size, and that the precision of the calculations is fixed and can be considered as a constant.

A computational advantage of signatures over histograms is that distributions defined in high-dimensional feature spaces can be matched more efficiently. This is because the only computational factor is the number of significant clusters in the distributions and not the dimension of the underlying space, although sometimes the two correlate. The EMD is also insensitive to the clustering algorithm that is used to find the significant clusters. If a cluster is split into smaller fragments, the EMD will consider them as similar as long as the distance between the fragments is small.

While the EMD works very well on signatures, it should not, in general, be applied to histograms. When the histograms are coarse, the bins are large, and the centers of neighboring bins cannot be considered as being close to each other, which causes the EMD to lose its main advantage of looking at cross-bin dissimilarities. On the other hand, computing the EMD on fine histograms can be too slow.

3.1.   One-Dimensional EMD

When the feature space is one-dimensional, and the ground distance is the Euclidean distance, and the two signatures have equal weight, the EMD can be computed fast without having to solve a linear optimization problem, as in the multi-dimensional case. In fact, the minimum cost distance between two one-dimensional distributions $f(t)$ and $g(t)$ is known to be the L_1 distance between the cumulative distribution functions [105, 11]

$$\int_{-\infty}^{\infty} \left| \int_{-\infty}^{x} f(t)dt - \int_{-\infty}^{x} g(t)dt \right| dx . \tag{2.2}$$

When the distributions are represented by histograms, Equation (2.2) is the match distance described in Chapter 1. When representing the distributions by signatures, the distance can be computed as follows: Let $P = \{(p_1, w_{p_1}), \dots, (p_m, w_{p_m})\}$ and $Q = \{(q_1, w_{q_1}), \dots, (q_n, w_{q_n})\}$ be two one-dimensional signatures.

THEOREM 2.2 *If the following conditions hold:*

1 *the feature space is one-dimensional,*

2 *the ground distance is* $d(p_i, q_j) = |p_i - q_j|,$

3 *the total weights of the two signatures are equal,*

then

$$EMD(P, Q) = \sum_{k=1}^{m+n-1} |\hat{p}_k - \hat{q}_k|(r_{k+1} - r_k) ,$$

where $r_1, r_2, \ldots, r_{m+n}$ is the sorted list $p_1, p_2, \ldots, p_m, q_1, q_2, \ldots, q_n$, and

$$\hat{p}_k = \sum_{i=1}^{m}[p_i \leq r_k]w_{p_i} \qquad \hat{q}_k = \sum_{j=1}^{n}[q_j \leq r_k]w_{q_j} \; ,$$

where $[\cdot]$ is 1 when its argument is true, and 0 otherwise.

Proof: The signatures P and Q can be written as the following one-dimensional continuous distribution functions:

$$p(t) = \sum_{i=1}^{m} w_{p_i}\delta(t - p_i) \qquad q(t) = \sum_{j=1}^{n} w_{q_j}\delta(t - q_j) \; ,$$

where δ is the Dirac delta function. Using Equation (2.2) we get

$$\text{EMD}(P, Q) = \int_{-\infty}^{\infty} \left| \int_{-\infty}^{x} p(t)dt - \int_{-\infty}^{x} q(t)dt \right| dx \; .$$

For $t < r_1$ and for $t > r_{m+n}$ we have $p(t) = q(t) = 0$ so we can write

$$\text{EMD}(P, Q) = \int_{r_1}^{r_{m+n}} \left| \int_{-\infty}^{x} p(t)dt - \int_{-\infty}^{x} q(t)dt \right| dx$$

$$= \int_{r_1}^{r_{m+n}} \left| \int_{-\infty}^{x} \sum_{i=1}^{m} w_{p_i}\delta(t - p_i)dt - \int_{-\infty}^{x} \sum_{j=1}^{n} w_{q_j}\delta(t - q_j)dt \right| dx$$

$$= \int_{r_1}^{r_{m+n}} \left| \sum_{i=1}^{m} w_{p_i} \int_{-\infty}^{x} \delta(t - p_i)dt - \sum_{j=1}^{n} w_{q_j} \int_{-\infty}^{x} \delta(t - q_j)dt \right| dx \; .$$

We can write the interval $[r_1, r_{m+n}]$ as $[r_1 - \epsilon, \; r_2 - \epsilon] \cup [r_2 - \epsilon, \; r_3 - \epsilon] \cup \ldots \cup [r_{m+n-1} - \epsilon, \; r_{m+n} - \epsilon] \cup [r_{m+n} - \epsilon, \; r_{m+n}] - [r_1 - \epsilon, \; r_1)$. The integrands in the following expression are the same for all the subintervals, so for simplicity, we specify it only once:

$$\text{EMD}(P, Q)$$

$$= \sum_{k=1}^{m+n-1} \int_{r_k - \epsilon}^{r_{k+1} - \epsilon} \left| \sum_{i=1}^{m} w_{p_i} \int_{-\infty}^{x} \delta(t - p_i)dt - \sum_{j=1}^{n} w_{q_j} \int_{-\infty}^{x} \delta(t - q_j)dt \right| dx$$

$$- \int_{r_1 - \epsilon}^{r_1} + \int_{r_{m+n} - \epsilon}^{r_{m+n}} \; .$$

The innermost integrals are equal to 1 only when $p_i < x$ (or $q_j < x$) and 0 otherwise. This occurs only for the subintervals that contain r_k where $p_i \leq r_k$

(or $q_j \leq r_k$), so

$$
\mathrm{EMD}(P,Q) = \sum_{k=1}^{m+n-1} \int_{r_k-\epsilon}^{r_{k+1}-\epsilon} \left| \sum_{i=1}^{m} w_{p_i} [p_i \leq r_k] - \sum_{j=1}^{n} w_{q_j} [q_j \leq r_k] \right| dx
$$
$$
- \int_{r_1-\epsilon}^{r_1} + \int_{r_{m+n}-\epsilon}^{r_{m+n}} .
$$

We now set $\epsilon \to 0$:

$$
\mathrm{EMD}(P,Q) = \sum_{k=1}^{m+n-1} \int_{r_k}^{r_{k+1}} \left| \sum_{i=1}^{m} w_{p_i} [p_i \leq r_k] - \sum_{j=1}^{n} w_{q_j} [q_j \leq r_k] \right| dx
$$
$$
= \sum_{k=1}^{m+n-1} \int_{r_k}^{r_{k+1}} |\hat{p}_k - \hat{q}_k| \, dx
$$
$$
= \sum_{k=1}^{m+n-1} |\hat{p}_k - \hat{q}_k| \, (r_{k+1} - r_k) .
$$

∎

3.2.   Lower Bounds

Retrieval speed can be increased if lower bounds to the EMD can be computed at a low expense. These bounds can significantly reduce the number of EMDs that actually need to be computed by prefiltering the database and ignoring images that are too far from the query. An easy-to-compute lower bound for the EMD between signatures with equal total weights is the distance between their centers of mass, as long as the ground distance is induced by a norm.

THEOREM 2.3 *Given signatures P and Q, let* $\mathbf{p}_i$ *and* $\mathbf{q}_j$ *be the coordinates of cluster i in the first signature, and cluster j in the second signature respectively. Then if*

$$
\sum_{i=1}^{m} w_{\mathbf{p}_i} = \sum_{j=1}^{n} w_{\mathbf{q}_j} , \qquad (2.3)
$$

then

$$
EMD(P,Q) \geq \|\bar{P} - \bar{Q}\| ,
$$

where $\| \cdot \|$ *is the norm that induces the ground distance, and* $\bar{P}$ *and* $\bar{Q}$ *are the centers of mass of P and Q respectively:*

$$
\bar{P} = \frac{\sum_{i=1}^{m} w_{\mathbf{p}_i} \mathbf{p}_i}{\sum_{i=1}^{m} w_{\mathbf{p}_i}} , \qquad \bar{Q} = \frac{\sum_{j=1}^{n} w_{\mathbf{q}_j} \mathbf{q}_j}{\sum_{j=1}^{n} w_{\mathbf{q}_j}} .
$$

Proof: Using the notation of equations (2.1a)-(2.1d),

$$\sum_{i=1}^{m}\sum_{j=1}^{n} d(\mathbf{p}_i, \mathbf{q}_j) f_{ij} = \sum_{i=1}^{m}\sum_{j=1}^{n} \|\mathbf{p}_i - \mathbf{q}_j\| f_{ij}$$

$$= \sum_{i=1}^{m}\sum_{j=1}^{n} \|f_{ij}(\mathbf{p}_i - \mathbf{q}_j)\| \qquad (f_{ij} \geq 0)$$

$$\geq \left\| \sum_{i=1}^{m}\sum_{j=1}^{n} f_{ij}(\mathbf{p}_i - \mathbf{q}_j) \right\|$$

$$= \left\| \sum_{i=1}^{m}\left(\sum_{j=1}^{n} f_{ij}\right)\mathbf{p}_i - \sum_{j=1}^{n}\left(\sum_{i=1}^{m} f_{ij}\right)\mathbf{q}_j \right\|$$

$$= \left\| \sum_{i=1}^{m} w_{\mathbf{p}_i}\mathbf{p}_i - \sum_{j=1}^{n} w_{\mathbf{q}_j}\mathbf{q}_j \right\|.$$

Dividing both sides by $\sum_{i=1}^{m}\sum_{j=1}^{n} f_{ij}$,

$$\frac{\sum_{i=1}^{m}\sum_{j=1}^{n} d(\mathbf{p}_i, \mathbf{q}_j) f_{ij}}{\sum_{i=1}^{m}\sum_{j=1}^{n} f_{ij}} \geq \left\| \frac{\sum_{i=1}^{m} w_{\mathbf{p}_i}\mathbf{p}_i - \sum_{j=1}^{n} w_{\mathbf{q}_j}\mathbf{q}_j}{\sum_{i=1}^{m}\sum_{j=1}^{n} f_{ij}} \right\|,$$

and using (2.1d) and (2.3) we get

$$\mathrm{EMD}(P, Q) \geq \left\| \frac{\sum_{i=1}^{m} w_{\mathbf{p}_i}\mathbf{p}_i}{\sum_{i=1}^{m} w_{\mathbf{p}_i}} - \frac{\sum_{j=1}^{n} w_{\mathbf{q}_j}\mathbf{q}_j}{\sum_{j=1}^{n} w_{\mathbf{q}_j}} \right\|$$

$$\geq \|\bar{P} - \bar{Q}\|.$$

∎

Using this lower bound in our color-based image retrieval system significantly reduced the number of EMD computations.

If the system is asked to return the n best matches, it checks all the images in the database, one at a time, and computes the lower bound distance between the image and the query. The EMD needs to be computed only if the lower bound distance is smaller than the distances between the n best matches so far and the query. Figure 2.3 shows the average number of EMD computations per query as a function of the number of images retrieved. This graph was generated by averaging over 200 random queries on an image database with 20,000 images using the color-based image retrieval system described in Chapter 3. The fewer images that are returned by a query, the smaller the lower bound distance has to be in order to compute the EMD, and the fewer EMDs that need to be computed

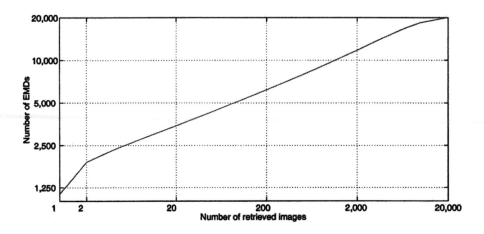

Figure 2.3. Log-log plot of the number of EMDs as a function of the number of images retrieved. The database contains 20,000 images.

due to our lower bound. This bound guarantees that no image is missed as a result of the saving in computation.

The lower bound defined in Theorem (2.3) holds only when the two signatures have equal weights and cannot be used for partial matching. In [11] two types of lower bounds are developed for the case of unequal weights: centroid-based and projection-based lower bounds. For completeness, we summarize here the main results.

The *CBOX (Centroid bounding BOX)* lower bound is an extension of the equal-weight centroid lower bound (Theorem (2.3)). Without loss of generality, assume that $w_P > w_Q$, where $w_P = \sum_{i=1}^m w_{\mathbf{p}_i}$ and $w_Q = \sum_{j=1}^n w_{\mathbf{q}_j}$. An optimal feasible flow matches an amount w_Q of the total weight of the heavier distribution P. If $C(P, w_Q)$ is the locus of all centroids of sub-distributions of P of total weight w_Q, then the minimum distance from the centroid of Q to $C(P, w_Q)$ is a lower bound on $\mathrm{EMD}(P, Q)$. The same statement holds with a bounding box $B(P, w_Q)$ of $C(P, w_Q)$ in place of $C(P, w_Q)$. The idea is to pre-compute (*i.e.* not at query time) a set of bounding boxes $B(P, w_Q)$ for $w_Q = 5\%, 10\%, 15\%, \ldots, 100\%$ of the weight of P. When a query Q with smaller weight comes in, the appropriate bounding box is chosen and the distance from the centroid of Q to this bounding box is computed (this is a constant time computation with respect to the number of points in P and Q).

The *projection-based* lower bounds are used when the ground distance is Euclidean. These bounds work with the projections of distributions onto lines through the origin. The Euclidean distance between two points is at least the distance between their projections onto a line through the origin, so the EMD between the projected distributions is a lower bound on the EMD between the original distributions (the weights in the projected distribution are the same as the weights in the original distribution). In the equal-weight case, computing the EMD between distributions along the line can be done much faster than in the general case. Since it follows from FeaSiBiLity conditions, the authors call this the *FSBL* lower bound.

The same technique, described in Section 3.1 for computing the EMD in one dimension, is modified to compute a lower bound on the EMD in one dimension between unequal-weight distributions. The idea is to record how much mass must flow over certain intervals if all the mass in the lighter distribution is to be matched. The projection-based lower bounds proposed in [11] are:

$$\text{PMAX}_{\text{FSBL}}(P, Q) = \max_{v \in V} \text{FSBL}(\text{proj}_v(P), \text{proj}_v(Q))$$

$$\text{PAMAX}_{\text{FSBL}}(P, Q) = \max_{e \in E} \text{FSBL}(\text{proj}_e(P), \text{proj}_e(Q))$$

$$\text{PASUM}_{\text{FSBL}}(P, Q) = \frac{1}{\sqrt{D}} \sum_{e \in E} \text{FSBL}(\text{proj}_e(P), \text{proj}_e(Q)) .$$

Here $\text{proj}_v(P)$ stands for the one-dimensional distribution which is the projection of P onto the line through the origin in direction v. V is a set of random directions, E is the set of feature space axes, and D is the dimension of the feature space. This lower bounds are named by different kinds of projections: Projection MAXimum (PMAX), Projection Axes MAXimum (PAMAX), and Projection Axes SUM (PASUM).

The hope with the projection-based bounds is that one could quickly tell that two distributions are very different by looking at one-dimensional projections, which are fast to compute using the results of Theorem (2.2). These often work well in cases when the CBOX lower bound does not (*i.e.* when one cannot tell the difference between distributions by looking at their averages).

More details on and experiments with these bounds can be found in [11].

4.   MORE ON THE GROUND DISTANCE

In Section 2 we claimed that correspondences between two distributions are essential to match perceptual similarity. The EMD indeed finds the best correspondences that minimize the total transportation cost; however, some of these correspondences might be between features that are far apart in the feature space, and therefore have large ground distances. This can have a big impact on the total cost, so that a few, very different, features can have a big effect on the distance between otherwise similar distributions.

Modifying the ground distance so that it saturates to some value limits the effect far features have on the EMD. An example of such a ground distance is

$$\hat{d}(\mathbf{p}_i, \mathbf{q}_j) = 1 - e^{-\alpha d(\mathbf{p}_i, \mathbf{q}_j)} , \tag{2.4}$$

where $d(\mathbf{p}_i, \mathbf{q}_j)$ is the non-saturated ground distance between clusters $\mathbf{p}_i$ and $\mathbf{q}_j$. The solid line in Figure 2.4 shows $\hat{d}$ as a function of d (in units of α^{-1}). The value for α should be chosen specifically for the feature space at hand so that it distinguishes between "close" and "far" distances in the feature space. In this book, we use Equation (2.4) as our ground distance, with

$$\alpha^{-1} = d(\mathbf{0}, \frac{1}{2}\sigma) ,$$

where $\mathbf{0}$ is the zero vector, $\sigma = [\sigma_1 \ldots \sigma_D]^T$ is a vector of standard deviations of the components of the features in each dimension from the overall distribution of all images in the database, and D is the dimensionality of the feature space. Assuming that the distribution of the features is unimodal, α is a measure of the spread of the distribution. The bigger the spread, the larger the distances are, in general.

The saturated ground distance, $\hat{d}(\mathbf{p}_i, \mathbf{q}_j)$, agrees with results from psychophysics. In [89] it is argued that the similarity between stimuli of any type can be expressed as *generalization data* by $g(d(S_i, S_j))$, where d is a perceptual distance between two stimuli, and g is a generalization function such as $g(d) = \exp(-d^\tau)$. This is equivalent to our dissimilarity measure which can be expressed in term of the similarity $g(d)$ by $1 - g(d)$, with $\tau = 1$.

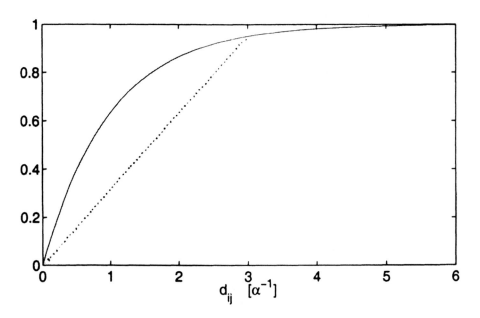

Figure 2.4. The saturated ground distances $\hat{d}$ (solid line), and $\tilde{d}$ (dotted line) as a function of the ground distance $d_{ij} = d(\mathbf{p}_i, \mathbf{q}_j)$. We assume that $\max_{i,j} d_{ij} = 3$.

THEOREM 2.4 *If the non-saturated ground distance, $d(\cdot, \cdot)$, is metric, then $\hat{d}(\cdot, \cdot)$ is also metric.*

Proof: Positive definiteness and symmetry clearly hold because $\alpha \geq 0$. We now prove that the triangle inequality holds, too. Given that $d(\mathbf{p}, \mathbf{q}) + d(\mathbf{q}, \mathbf{r}) \geq d(\mathbf{p}, \mathbf{r})$, and using positive definiteness, we can write

$$
\begin{aligned}
0 \; \leq \; & \hat{d}(\mathbf{p}, \mathbf{q}) \hat{d}(\mathbf{q}, \mathbf{r}) \\
= \; & \left(1 - e^{-\alpha d(\mathbf{p}, \mathbf{q})}\right) \left(1 - e^{-\alpha d(\mathbf{q}, \mathbf{r})}\right) \\
= \; & 1 - e^{-\alpha d(\mathbf{p}, \mathbf{q})} - e^{-\alpha d(\mathbf{q}, \mathbf{r})} + e^{-\alpha(d(\mathbf{p}, \mathbf{q}) + d(\mathbf{q}, \mathbf{r}))} \\
\leq \; & 1 - e^{-\alpha d(\mathbf{p}, \mathbf{q})} - e^{-\alpha d(\mathbf{q}, \mathbf{r})} + e^{-\alpha d(\mathbf{p}, \mathbf{r})} \\
= \; & \left(1 - e^{-\alpha d(\mathbf{p}, \mathbf{q})}\right) + \left(1 - e^{-\alpha d(\mathbf{q}, \mathbf{r})}\right) - \left(1 - e^{-\alpha d(\mathbf{p}, \mathbf{r})}\right) \\
= \; & \hat{d}(\mathbf{p}, \mathbf{q}) + \hat{d}(\mathbf{q}, \mathbf{r}) - \hat{d}(\mathbf{p}, \mathbf{r}) \, ,
\end{aligned}
$$

and hence, $\hat{d}(\mathbf{p}, \mathbf{q}) + \hat{d}(\mathbf{q}, \mathbf{r}) \geq \hat{d}(\mathbf{p}, \mathbf{r})$. ∎

COROLLARY 2.1 *The resulting EMD is metric for signatures of equal weights.*

Although metric, $\hat{d}$ is not induced by a norm, and therefore the lower bound defined in Section 3.2 is not valid for the EMD that uses $\hat{d}$ as its ground distance. We denote this EMD by $\widehat{\text{EMD}}$. However, a valid lower bound can be defined using an EMD that uses the following ground distance:

$$
\tilde{d}(\mathbf{p}_i, \mathbf{q}_j) = \frac{\max_{i,j} \hat{d}(\mathbf{p}_i, \mathbf{q}_j)}{\max_{i,j} d(\mathbf{p}_i, \mathbf{q}_j)} d(\mathbf{p}_i, \mathbf{q}_j) \, .
$$

We denote the resulting EMD by $\widetilde{\text{EMD}}$. This ground distance is a scaled Euclidean distance and therefore it is induced by a norm.

THEOREM 2.5

$$
\widehat{\text{EMD}}(P, Q) \geq \widetilde{\text{EMD}}(P, Q) \geq \frac{\max_{i,j} \hat{d}(\mathbf{p}_i, \mathbf{q}_j)}{\max_{i,j} d(\mathbf{p}_i, \mathbf{q}_j)} \|\bar{P} - \bar{Q}\| \, .
$$

Proof: The first part, $\widehat{\text{EMD}}(P, Q) \geq \widetilde{\text{EMD}}(P, Q)$, holds because $\hat{d}(\mathbf{p}_i, \mathbf{q}_j)$ is convex, and therefore $\hat{d}(\mathbf{p}_i, \mathbf{q}_j) \geq \tilde{d}(\mathbf{p}_i, \mathbf{q}_j)$ for all possible i and j, as shown in Figure 2.4. The second part, $\widetilde{\text{EMD}}(P, Q) \geq \frac{\max_{i,j} \hat{d}(\mathbf{p}_i, \mathbf{q}_j)}{\max_{i,j} d(\mathbf{p}_i, \mathbf{q}_j)} \|\bar{P} - \bar{Q}\|$, results directly from the lower bound defined in Section 3.2. ∎

5. EXTENSIONS

A few extensions were developed for the EMD [12]. For completeness we now describe these extensions.

5.1.   The Partial EMD (EMD$^{\gamma}$)

The partial EMD forces only some given fraction $0 < \gamma \leq 1$ of the weight of the lighter distribution to be matched. The constraint (2.1d) for the usual EMD is replaced by

$$\sum_{i=1}^{m}\sum_{j=1}^{n} f_{ij} = \gamma \min(\sum_{i=1}^{m} w_{\mathbf{p}_i}, \sum_{j=1}^{n} w_{\mathbf{q}_j}) ,$$

and the minimum work is normalized by $\gamma \min(\sum_{i=1}^{m} w_{\mathbf{p}_i}, \sum_{j=1}^{n} w_{\mathbf{q}_j})$. The partial EMD makes the EMD robust to a fraction $1 - \gamma$ missing data/outliers.

5.2.   The Restricted EMD (τ-EMD)

The τ-EMD measures the minimum fraction of mass that cannot be matched when flows are restricted to at most τ ground distance units. The extreme values are τ-EMD $= 0$ when all the mass can be matched, and τ-EMD $= 1$ when none of the mass can be matched. The definition is

$$\tau\text{-EMD}(P,Q) = 1 - \frac{\max_F \sum_{i,j} f_{ij}[d(\mathbf{p}_i, \mathbf{q}_j) \leq \tau]}{\min(\sum_{i=1}^{m} w_{\mathbf{p}_i}, \sum_{j=1}^{n} w_{\mathbf{q}_j})} .$$

Here $[d(\mathbf{p}_i, \mathbf{q}_j) \leq \tau]$ equals 1 if $d(\mathbf{p}_i, \mathbf{q}_j) \leq \tau$ and 0 otherwise. The numerator measures the maximum amount of mass that can be matched. The set of feasible flows over which the optimization is computed is the same as for the regular EMD.

The τ-EMD tells what percentage of the features matches well (if τ is set low), instead of averaging all features, including the bad matches (with correspondences over large distances).

6.   SUMMARY

The Earth Mover's Distance is a general and flexible metric and has desirable properties for image retrieval. It allows for partial matches, and it can be applied to variable-length representations of distributions. Lower bounds are readily available for it, and it can be computed efficiently when the signatures are not too large. The EMD works better when applied to signatures, than to global histograms. Histograms with few bins invalidate the ground distances, while EMDs on histograms with many bins are slow to compute. Because of the advantages of the EMD, we believe that the EMD can be of use both for understanding vision problems related to distributions, as exemplified in the next two chapters with color and texture, and as a fundamental element of image retrieval systems.

Chapter 3

COLOR-BASED IMAGE SIMILARITY

One should absorb the color of life, but one should never remember its detail.
—Oscar Wilde, 1854–1900

Color plays an important role in content-based image retrieval. In this chapter we define color features in an appropriate, perceptually uniform color space, where color distributions describe the contents of entire images. Summarizing these distributions by color signatures and using the Earth Mover's Distance leads to a powerful color-based retrieval paradigm. Combining the color information of the pixels, with their positions in the image leads to a distance measure where not only the color contents matters but also the layout of the color in the image.

1. COLOR FEATURES

Human color perception is based on the incidence of visible light (with wavelengths in the range of 400 nm to 700 nm) upon the retina. Since there are three types of color photoreceptor cone cells in the retina, each with a different spectral response curve, all colors can be completely described by three numbers, corresponding to the outputs of the cone cells. In 1931, the Commission Internationale de l'Éclairage (CIE) adopted standard curves for the color photoreceptor cone cells of a hypothetical standard observer, and defined the CIE XYZ tristimulus values, where all visible colors can be represented using only positive values of X, Y and Z.

1.1. Color Spaces

Color spaces are used to specify, create and visualize color information. While the CIE XYZ space describes all the colors we can perceive, other color spaces are subsets of this space and represent fewer colors than we can see. For instance, the RGB color space, as used by television displays, can be visualized as a cube with red, green and blue axes. Different applications have different needs which can be handled better using different color spaces. Many color spaces have been developed including the following (see [106] for more information):

RGB (Red-Green-Blue) An additive color system, based on adding the three primary colors. RGB is commonly used by CRT displays where proportions of excitation of red, green and blue emitting phosphors produce colors when visually fused.

CMY(K) (Cyan-Magenta-Yellow(-Black)) A subtractive color system commonly used in printing. The black component is redundant and is used for technical reasons such as improving the appearance of the black color and reducing costs.

HSL (Hue-Saturation-Lightness) and HSB (Hue-Saturation-Brightness) Intuitive spaces that allows users to specify colors easily. Separating the luminance component has advantages in certain image processing applications.

YIQ, YUV, YCbCr Used for different standards of television transmission (NTSC, PAL, and digital TV respectively).

Opponent Colors Used for modeling color perception. It is based on the fact that for color perception, some pairs of hues cannot coexist in a single color sensation (e.g. red and green) [103].

1.2. Perceptually Uniform Color Spaces

For an image retrieval system, it is important to be able to measure differences between colors in a way that matches perceptual similarity as well as possible. This task is simplified by the use of *perceptually uniform* color spaces. A color space is perceptually uniform if a small perturbation of a color will produce the same change in perception anywhere in the color space. In 1976, the CIE standardized two perceptually uniform color spaces (since no single system could be agreed upon), $L^*u^*v^*$ (CIELuv) and $L^*a^*b^*$ (CIELab). The L^* component defines the luminance, and the two other components (u^*, v^* and a^*, b^*) define the chrominance. Both spaces are defined with respect to the CIE XYZ color space, using a *reference white* $[X_n \ Y_n \ Z_n]^T$. Following ITU-R Recommendation BT.709, we use D_{65} as the reference white so that $[X_n \ Y_n \ Z_n] = [0.9504511.088754]$ (see [70]). In practice, most images are represented in the RGB color space. The conversions between RGB and CIE-XYZ are linear:

$$\begin{bmatrix} X \\ Y \\ Z \end{bmatrix} = \begin{bmatrix} 0.412453 & 0.357580 & 0.180423 \\ 0.212671 & 0.715160 & 0.072169 \\ 0.019334 & 0.119193 & 0.950227 \end{bmatrix} \begin{bmatrix} R \\ G \\ B \end{bmatrix}$$

and

$$\begin{bmatrix} R \\ G \\ B \end{bmatrix} = \begin{bmatrix} 3.240479 & -1.537150 & -0.498535 \\ -0.969256 & 1.875992 & 0.041556 \\ 0.055648 & -0.204043 & 1.057311 \end{bmatrix} \begin{bmatrix} X \\ Y \\ Z \end{bmatrix}$$

For both the CIELuv and CIELab, the Euclidean distance is used to compute the distance between (close) colors. In the following we give the transformations from CIE XYZ:

The luminance is defined similarly for both spaces,

$$L^* = \begin{cases} 116 \, (Y/Y_n)^{1/3} - 16 & \text{if } Y/Y_n > 0.008856 \\ 903.3(Y/Y_n) & \text{otherwise} \end{cases},$$

while the chrominances are defined differently for CIELuv and for CIELab.

1.2.1 CIELuv

$$\begin{aligned} u^* &= 13L^*(u' - u'_n), \\ v^* &= 13L^*(v' - v'_n), \end{aligned}$$

where

$$\begin{aligned} u' &= \frac{4X}{X + 15Y + 3Z}, \\ v' &= \frac{9Y}{X + 15Y + 3Z}, \end{aligned}$$

and u'_n and v'_n have the same definitions for u' and v' but are computed using X_n, Y_n, and Z_n.

1.2.2 CIELab

$$\begin{aligned} a^* &= 500 \, (f(X/X_n) - f(Y/Y_n)), \\ b^* &= 200 \, (f(Y/Y_n) - f(Z/Z_n)), \end{aligned}$$

where

$$f(t) = \begin{cases} t^{1/3} & \text{if } Y/Y_n > 0.008856, \\ 7.787t + 16/116 & \text{otherwise}. \end{cases}$$

1.2.3   Spatial Extension

So far, we assumed that the perception of a single color is independent of colors that surround it. Is general, once the visual angle of a region in a color image gets smaller than 2 degrees, this assumption does not hold. To account for the spatial effect of color perception the S-CIELab[107] was developed as a spatial extension to the CIELab color space. A color image is first transformed into the opponent-colors space. Each of the three component images is convolved with a kernel whose shape is determined by the visual spatial sensitivity in that color band. Finally, the filtered representation is transformed to CIELab.

We found that using CIELab for color-based image retrieval results in better agreement with color perception than when using RGB. This is because RGB is not perceptually uniform, so that small perceptual differences between some pairs of colors are equivalent to larger perceptual differences in other pairs.

2.   COLOR SIGNATURES

To compute the color signature of a image, we first smooth each band of the image's RGB representation slightly to reduce possible color quantization and dithering artifacts. We then transform the image into S-CIELab. At this point each image can be conceived as a distribution of points in CIELab, where a point corresponds to a pixel in the image. We coalesce this distribution into clusters of similar colors (25 units in any of the L^*, a^*, b^* axes). Because of the large number of images to be processed in typical database applications, clustering must be performed efficiently. To this end, we devised a novel two-stage algorithm based on a k-d tree [5]. In the first phase, approximate clusters are found by excessive subdivisions of the space into rectangular blocks, splitting each block in turn in the middle of its longest dimension, and stopping when all sides of the blocks become smaller than a minimum allowed cluster size (12.5 units). Since, because of our simple splitting rule, clusters might be split over a few blocks, we use a second phase to recombine them. This is done by performing the same clustering procedure using the cluster centroids from the first phase, after shifting the space coordinates by one-half the minimum block size (6.25 units). Each new cluster contributes a pair $(\mathbf{p}, w_{\mathbf{p}})$ to the signature representation of the image, where $\mathbf{p}$ is the mean of the cluster, and $w_{\mathbf{p}}$ is the fraction of image pixels in that cluster. At this point, for compactness, we remove clusters with insignificant weights (less than 0.1%). In our database of 20,000 color images from the Corel Stock Photo Library, the average signature has 8.8 clusters. An example of a color signature is given in Figure 3.1. Part (b) shows the color signature of the image in part (a). In part (c), the image was rendered using only the colors from the signature. We can see that, in this example, the color information is well retained.

(a)   (b)   (c)

Figure 3.1. An example of a color signature. (a) Original image. (b) Color signature. (c) Rendering the image using the signature colors. *(This is a color figure)*

3. COLOR-BASED IMAGE RETRIEVAL

After the color signatures are computed for all the images in the database, it is possible to retrieve the most similar images to a given query. In our system, color-based query is based on a color signature that is the result of one of three basic types of queries:

1 The signature describes the content of an entire image (*query-by-example*). The image can be part of the database or come from other sources. The total weight of the signature, in this case, is the same as for the other signatures in the database.

2 The users describe the colors they are looking for either explicitly, by specifying desired colors and their amounts, or implicitly, by drawing a sketch (*query-by-sketch*) from which a signature is inferred. Now the total weight of the query's signature can be less than for the other signatures in the database, resulting in a partial query. The weight difference can match any color and is referred to as "don't care."

3 The signature is generated by the system as a result of some user action. In Chapter 7 we describe a navigation scheme where the signatures are generated automatically.

Once the query is represented by a color signature, the retrieval system returns the images in the database whose signatures have minimum distance from the query.

To compute the distance between color signatures, we use the Earth Mover's Distance, with the Euclidean distance as a natural choice of ground distance. We performed our color-based image retrieval on a collection of 20,000 color images. Figures 3.2 and 3.3 show examples of color-based image retrieval. In the first three examples, the color content of an image was used as the query, while in the last example a partial query was used: The user looked for flowers and specified the query: "Find images with 20% pink, 40% green, and the rest can be any color." The results is these examples are good because there is a high correlation between the semantic queries and the colors used to describe them. A more systematic evaluation of the EMD for color-based retrieval and a comparison with other dissimilarity measures are given in Chapter 5.

Color information is usually not sufficient for semantic image retrieval. When the query cannot be well described in terms of color, there is a discrepancy between the semantic query and the syntactic features (color), and the query is bound to fail. An example is given in Figure 3.4 where an image of a woman was given as the query. Here the system responded with an image of a skunk as the best match. Although the color contents of the two images are very similar, they are obviously very different semantically.

4. JOINT DISTRIBUTION OF COLOR AND POSITION

In many cases, global color distributions that ignore the positions of the colors in the image are not sufficient for good retrieval. For example, consider the following two color images: In the first, there are blue skies *above* a green field, while in the other there is a blue lake *below* green plants. Although the color distributions might be very similar, the positions of the colors in the image are very different and may have to be taken into account by the query. This can be achieved by modifying the color distance in Section 3 as follows: Instead of using the three-dimensional CIELab color space, we use a five-dimensional space whose first three dimensions are the CIELab color space, and the other two are the (x, y) position of each pixel. We normalize the image coordinates for different image sizes and aspect ratios and use the same clustering algorithm as before. The average signature size in our 20,000 image database is now 18.5.

The ground distance is now defined as

$$\left[(\Delta L)^2 + (\Delta a)^2 + (\Delta b)^2 + \lambda \left((\Delta x)^2 + (\Delta y)^2 \right) \right]^{\frac{1}{2}}.$$

The parameter λ defines the importance of the color positions relative to the color values. Figure 3.5 shows the effect of position information. Part (a) shows the best matches to the query shown on the left which specifies 20% green at the top of the image, 20% blue at the bottom, and 60% "don't care" everywhere

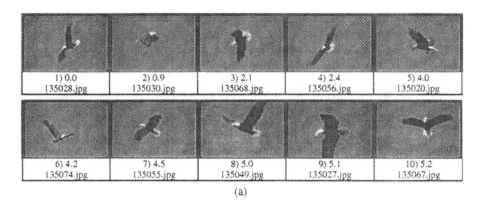

(a)

(b)

Figure 3.2. Examples of color-based image retrieval. The first image was used as the query. (a) Searching for eagles. (b) Searching for elephants. Using only color, elephants and hippopotamuses are very similar. *(This is a color figure)*

else. The typical returned images contain blue water at the bottom and green plants at the top. In part (b) the positions of the blue and green are exchanged. Now the returned images show blue skies on top of green fields.

Another example is shown in Figure 3.6 where the leftmost image of a skier was used as the query. λ can be found experimentally, or by using the method in [93]. Part (a) shows the 6 best matches when position information was ignored ($\lambda = 0$). Part (b) uses position information ($\lambda = 0.5$). Exact color matches are somewhat compromised in order to get more similar positional layouts.

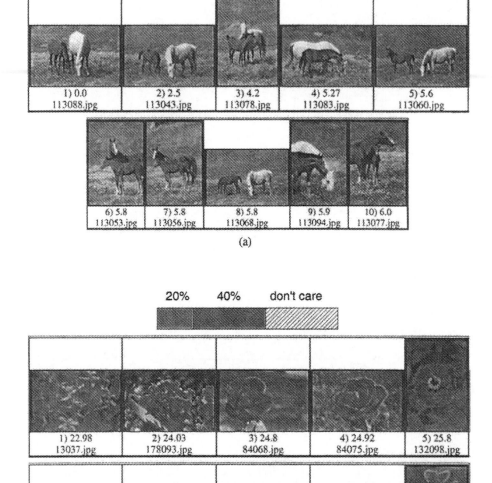

Figure 3.3. More examples. (a) Searching for horses. (b) Searching for flowers using a partial query, shown at the top. *(This is a color figure)*

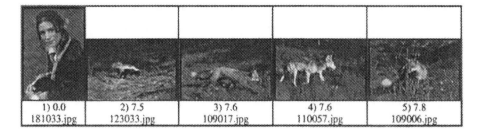

1) 0.0 181033.jpg	2) 7.5 123033.jpg	3) 7.6 109017.jpg	4) 7.6 110057.jpg	5) 7.8 109006.jpg

Figure 3.4. A bad example. This is due to the discrepancy between the semantic query ("beautiful woman") and the syntactic features (the color content). *(This is a color figure)*

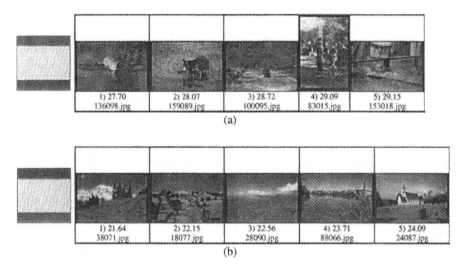

1) 27.70 136098.jpg	2) 28.07 159089.jpg	3) 28.72 100095.jpg	4) 29.09 83015.jpg	5) 29.15 153018.jpg

(a)

1) 21.64 38071.jpg	2) 22.15 18077.jpg	3) 22.56 28090.jpg	4) 23.71 88066.jpg	5) 24.09 24087.jpg

(b)

Figure 3.5. Using position information (the queries are on the left). (a) 20% green at the top and 20% blue at the bottom. (b) 20% blue at the top and 20% green at the bottom. *(This is a color figure)*

5. SUMMARY

In this chapter we showed that by carefully choosing a color space together with a perceptual appropriate ground distance, the EMD is a good measure for dissimilarity between color images. We showed retrieval examples over a 20,000-image database, using query-by-example and partial queries. We also presented an extension that in addition to the color information itself, also uses the absolute position of the colors in the images. In the next chapter we present measures for texture-based image similarity, using similar tools to those used in this chapter.

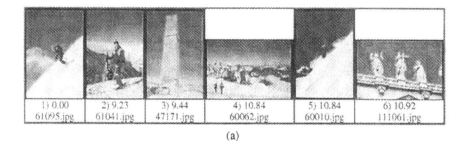

Figure 3.6.   Using the leftmost image of a skier as the query. The six best matches without position information (a) and with position information (b). *(This is a color figure)*

Chapter 4

TEXTURE-BASED IMAGE SIMILARITY

The imagination is of so delicate a texture that even words wound it.
—Hazlitt, 1778–1830

Although the notion of visual texture seems, at first, to be familiar and in-tuitive, there is no definition of texture that is universally accepted. Texture usually refers to a repetition of a perceptually similar pattern over an image region, such that the region is perceived as homogeneous by a human observer. Often, for structural textures (in contrast to statistical ones) such a pattern is referred to as the "texton". A pattern and its repetition can often be represented well by a spectral decomposition of the texture which captures the amount of repetition in different scales and in different orientations. In this chapter we use Gabor filters to represent texture. We develop two classes of dissimilarity measures: for homogeneous textures and for images that contain many textures, such as natural images. Finally, we present a preprocessing method to enhance the description of the texture content of images.

1. TEXTURE FEATURES

While color is a purely pixel-wise property of images, texture involves a notion of spatial extent: a single point has no texture. To account for spatial neighborhoods, a texture is commonly computed by the projection of the image intensity function onto an appropriate basis of functions. This is referred to as a *spectral decomposition*, because each of the different basis functions is usually concentrated in a different area of the two-dimensional spatial frequency domain. In such a representation, a texture is represented by a vector of values, each corresponding to the energy in a specified scale and orientation subband. Spectral decomposition methods include using quadrature filters [47], Gabor filters [7, 25, 6, 56], oriented derivatives of a Gaussian [67, 36], and the cortex transform [104].

Gabor filters [30] have been shown to give good results in comparison to other texture decompositions [57]. There is strong evidence that simple cells in the primary visual cortex can be modeled by Gabor functions tuned to detect different orientations and scales on a log-polar grid [20]. We therefore chose to employ such a set of Gabor functions to characterize texture.

1.1.   Gabor Filters

Gabor filters can be seen as two-dimensional wavelets [19]. When applied on an image I, the discretization of a two-dimensional wavelet is given by

$$W_{mlpq} = \iint I(x, y) \psi_{ml} (x - p\Delta x, y - q\Delta y)\, dx dy \,,$$

where $\Delta x, \Delta y$ is the spatial sampling rectangle (we use $\Delta x = \Delta y = 1$), p, q are image position, and m and l specify the scale and orientation of the wavelet respectively, with $m = 0, \ldots, M - 1$ and $l = 0, \ldots, L - 1$. The notation

$$\psi_{ml}(x, y) = a^{-m} \psi(\tilde{x}, \tilde{y}) \,, \tag{4.1}$$

where

$$
\begin{aligned}
\tilde{x} &= a^{-m}(x \cos\theta + y \sin\theta) \,, \\
\tilde{y} &= a^{-m}(-x \sin\theta + y \cos\theta) \,,
\end{aligned}
$$

denotes a dilation of the *mother wavelet* $\psi(x, y)$ by a^{-m} (a is the scale parameter), and a rotation by $\theta = l\Delta\theta$, where $\Delta\theta = 2\pi/L$ is the orientation sampling period.

$\psi_{ml}(x, y)$ is defined so that all the wavelet filters have the same energy:

$$
\begin{aligned}
\iint |\psi_{ml}(x, y)|^2 dx dy &= \iint |a^{-m} \psi(\tilde{x}, \tilde{y})|^2 dx dy \\
&= \iint |a^{-m} \psi(\tilde{x}, \tilde{y})|^2 \, |J^{-1}| d\tilde{x} d\tilde{y} \\
&= \iint |a^{-m} \psi(\tilde{x}, \tilde{y})|^2 a^{2m} d\tilde{x} d\tilde{y} \\
&= \iint |\psi(\tilde{x}, \tilde{y})|^2 d\tilde{x} d\tilde{y} \,, \tag{4.2}
\end{aligned}
$$

which is obviously independent of the choice of m and l. Here J is the Jacobian

$$
J = \begin{bmatrix} a^{-m} \cos\theta & a^{-m} \sin\theta \\ -a^{-m} \sin\theta & a^{-m} \cos\theta \end{bmatrix} \,.
$$

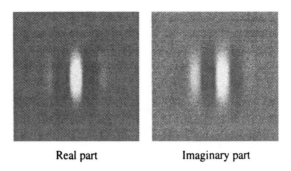

Real part   Imaginary part

Figure 4.1.   The components of $\psi(x, y)$.

We use this family of wavelets as our filter bank with the following Gabor function as the mother wavelet:

$$\psi(x, y) = \frac{1}{2\pi\sigma_x\sigma_y} \exp\left(-\frac{1}{2}\left(\frac{x^2}{\sigma_x^2} + \frac{y^2}{\sigma_y^2}\right)\right) \exp(i2\pi W x) .$$

The constant W determines the frequency bandwidth of the filters. We use $W = 0.5$, which corresponds to a half-amplitude bandwidth of 1 octave and is consistent with neurophysiological findings [52]. The real and imaginary parts of $\psi(x, y)$ are shown in Figure 4.1.

A Gabor function is essentially a two-dimensional Gaussian modulated with a complex exponential; therefore its frequency domain representation is a two-dimensional Gaussian with appropriate displacement along the u-axis, where we use (u, v) to index the frequency domain

$$\begin{aligned} \Psi(u, v) &= e^{-2\pi^2(\sigma_x^2 u^2 + \sigma_y^2 v^2)} ** \delta(u - W) \\ &= \exp(-2\pi^2(\sigma_x^2(u - W)^2 + \sigma_y^2 v^2)) \\ &= \exp\left(-\frac{1}{2}\left(\frac{(u - W)^2}{\sigma_u^2} + \frac{v^2}{\sigma_v^2}\right)\right) , \end{aligned} \qquad (4.3)$$

where $\sigma_u = (2\pi\sigma_x)^{-1}$ and $\sigma_v = (2\pi\sigma_y)^{-1}$. We use $**$ to represent two-dimensional convolution.

Filters with the same energy, as in Equation (4.2), are desirable for many applications and are commonly used for texture analysis. However, they are undesirable for computing the texture dissimilarity of natural images. It was shown that in natural images the amplitude spectrum typically falls off as the reciprocal of frequency [27], which results in smaller responses for high frequency Gabor filters. This can be fixed by removing the factor a^{-m} (which corresponds to 1/frequency) from Equation (4.1), resulting in

$$\hat{\psi}_{ml}(x, y) = \psi(\tilde{x}, \tilde{y}) ,$$

and our two-dimensional wavelet becomes

$$\hat{W}_{mlpq} = \iint I(x,y)\hat{\psi}_{ml}\left(x - p\Delta x, y - q\Delta y\right) dx dy .$$

For natural images, similar average amounts of energy will now be captured by all filters in all scales, providing a normalized feature space.

1.2. Filter Bank Design

We now design a bank of Gabor filters that tile the frequency space so that an image texture will be well represented by the set of individual filter responses. Our design strategy follows similar principles as in [6, 54]:

1 *Uniform separation in orientation.* Assuming rotational symmetry, all filters in a specific scale should have the same angular standard deviation (σ_v) and should be equally spaced in the orientation axis.

2 *Exponential separation in scale.* The widths of the filters should increase exponentially with distance from the center of the (u, v)-plane. This follows our intuitive notion of scale (a difference of one scale refers to multiplication by the scale parameter).

3 *Continuous coverage of the frequency space.* We design the filters such that the half-width contours of two neighboring filters touch each other both in the scale and in the orientation axes.

Let U_l and U_h be the lowest and highest frequencies of interest, such that the coarsest-scale filter and the finest-scale filter are centered in the frequency domain at distances U_l and U_h from the origin, respectively. If M is the number of scales, the filter centers, q_m, are spaced in exponentially increasing distances:

$$q_m = a^m U_l \qquad m = 0, 1, 2, \ldots, M - 1 .$$

The scale parameter a is defined by the constraint $U_h = a^{M-1} U_l$:

$$a = \left(\frac{U_h}{U_l}\right)^{\frac{1}{M-1}} .$$

The standard deviations of the Gabor filters, σ_u and σ_v, which are used in Equation 4.3, are

$$\sigma_u = \frac{a-1}{a+1} \frac{U_h}{\sqrt{2\ln 2}} ,$$

$$\sigma_v = \tan\left(\frac{\pi}{2L}\right) \sqrt{\frac{U_h^2}{2\ln 2} - \sigma_u^2} ,$$

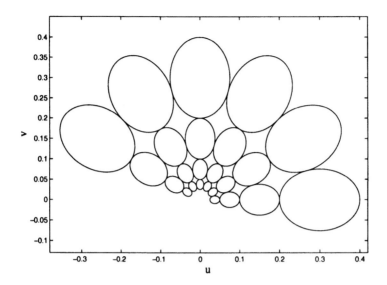

Figure 4.2. The half-amplitude of Gabor filters in the frequency domain using four scales and six orientations. Here we use $U_h = 0.3$ and $U_l = 0.04$.

as derived in Appendix A. For our system we use $U_h = 0.3$ and $U_l = 0.04$ (the upper bound for the spatial frequency is 0.5, which is the Nyquist frequency for an image that is sampled at every pixel). Figure 4.2 shows the location and width of the resulting Gabor filters, using four scales and six orientations.

Applying the Gabor filter bank to an image results for every image pixel (p, q) in a M by L array of responses to the filter bank. We retain only the magnitudes of the responses:

$$F_{mlpq} = |\hat{W}_{mlpq}| \quad m = 0, \ldots, M-1 \quad l = 0, \ldots, L-1 .$$

as the magnitudes encode the energy content and are independent of position within a texture.

This array of numbers is our *texture feature*. Figure 4.3 shows two examples of texture features using five scales and eight orientations. Darker squares represent stronger responses. The top texture has one dominant orientation in a fine scale, while the bottom texture has three dominant orientations is a coarser scale. The magnitude of each texture's discrete Fourier transform (DFT) is shown in the middle column of the figure.

2. HOMOGENEOUS TEXTURES

In this section we present dissimilarity measures that compute distances between homogeneous textures. Homogeneity allows us to assume that all of

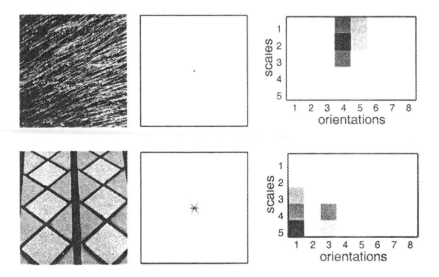

Figure 4.3. Texture features. *Left:* Patches from fabric (top) and tile (bottom) textures. *Middle:* DFT magnitude. *Right:* Texture features; the horizontal and vertical axes correspond to orientations and scale respectively.

the texture features are similar; therefore, we can represent textures by the mean of the texture descriptors over the entire texture. We use the energies of the Gabor filter responses:

$$E_{ml} = \frac{\sum_{p,q} F_{mlpq}^2}{\sum_{m,l,p,q} F_{mlpq}^2} \,.$$

The denominator normalizes E_{ml} so that $\sum_{m,l} E_{ml} = 1$. This normalization enhances texture content and reduces the effects of illumination and contrast. With this texture representation, the texture space can be seen as a unit hypersphere in a M-by-L dimensional space and a specific texture as a unit vector in that space.

The notion of texture dissimilarity varies with the task at hand. While for texture classification one may want the dissimilarity to be invariant to rotations of the texture and perhaps also to changes in scale, for texture segmentation these invariants may be inappropriate. In this section we show how to use the EMD to define a family of distances between textures. We define three distances: a distance with no invariance, a rotation-invariant distance and a distance with both rotation and scale invariance.

The homogeneous texture signature in all the distances defined in this section is the texture vector $\mathbf{E} = \{E_{ml}\}$ itself, where every entry E_{ml} corresponds to a cluster in the two-dimensional, log-polar space, positioned at (m, l) with the value of E_{ml} being the weight of the cluster. Since for the computation of the EMD the two signatures do not have to be of the same size, the signatures can be compacted by removing clusters with zero or very small weights, which have very small influence on the result of the EMD. This significantly reduces the size of the EMD problem which can be computed much faster. In practice, most of the bins for most textures can be pruned with negligible change in the EMD.

2.1. No Invariance

We can use the EMD as a distance measure between two homogeneous texture signatures. To this end, we define our ground distance to be the L_1 distance in log-polar space (similar results were obtained with the Euclidean distance). Since log-polar space is cylindrical, we have two possible distances between a pair of points. We define the ground distance to be the shorter of the two distances. The ground distance between the points (m_1, l_1) and (m_2, l_2) is therefore:

$$d\left((m_1, l_1), (m_2, l_2)\right) = |\Delta l| + \alpha |\Delta m| , \qquad (4.4)$$

where

$$\Delta m = m_1 - m_2 \quad , \quad \Delta l = \min(|l_1 - l_2|, \ L - |l_1 - l_2|) .$$

The parameter α controls the relative importance of scale versus orientation. We used $\alpha = 1$ with good results. Other choices of α can result from applications or from psychophysical experiments. Our ground distance is metric (see proof in Section 2.4), and therefore the defined EMD is metric as well, as shown in Section 2.

2.2. Rotation Invariance

In our log-polar array, rotation is equivalent to a cyclic shift along the orientation axis. Although inefficient, we can achieve rotation invariance by an exhaustive search for the minimal distance over all possible shifts in orientation. An algorithm that can avoid this exhaustive search was developed in [12], based on an Expectation-Maximization (EM) type optimization that is used to compute the EMD under transformations. This algorithm can significantly reduce the computation times to compute the rotation-invariant distance and the rotation- and scale-invariant distance defined later.

Let $\mathbf{E}_1$ and $\mathbf{E}_2$ be two homogeneous texture signatures. An EMD that is invariant to texture rotation is

$$\mathrm{EMD}(\mathbf{E}_1, \mathbf{E}_2) = \min_{l_s = 0, \dots, L-1} \mathrm{EMD}(\mathbf{E}_1, \mathbf{E}_2, l_s) ,$$

where $\mathrm{EMD}(\mathbf{E}_1, \mathbf{E}_2, l_s)$ is the EMD with orientation shift l_s. The ground distance in Equation (4.4) uses the same Δm but

$$\Delta l = \min(\ |l_1 - l_2 + l_s \pmod{L}|, L - |l_1 - l_2 + l_s \pmod{L}|) .$$

2.3.   Rotation and Scale Invariance

Scale invariance can be obtained in a similar manner. In the log-polar array, scale invariance can be seen as invariance to shifts in the scale axis. An EMD that is invariant to both rotation and scale is

$$\text{EMD}(\mathbf{E}_1, \mathbf{E}_2) = \min_{\substack{l_s = 0, \dots, L-1 \\ m_s = -(M-1), \dots, M-1}} \text{EMD}(\mathbf{E}_1, \mathbf{E}_2, l_s, m_s) ,$$

where $\text{EMD}(\mathbf{E}_1, \mathbf{E}_2, l_s, m_s)$ is the EMD with orientation shift l_s and scale shift m_s. The ground distance is similar to the rotation-invariant case with

$$\Delta m = m_1 - m_2 + m_s .$$

2.4.   Proving that the Distances are Metric

Here we prove that all distances defined in this section for homogeneous textures are indeed metric. Non-negativity and symmetry hold trivially in all cases, so we only need to prove that the triangle inequality holds. In Section 2 it is proven that if the ground distance is metric, the EMD is metric, so to prove that the non-invariant distance is metric, we need to show that the ground distance (L_1 in our case) we use in the cylindrical log-polar space is metric. The projection of two points A and C on the cylinder onto its circular base divides it into two circular arcs (in the degenerate case one arc can be a point). The projection of a third point B can be in only one of these two arcs. Now the cylinder can be unfolded into a rectangle by a vertical cut anywhere through the other arc. Since in this two-dimensional plane $d(A, C) \leq d(A, B) + d(B, C)$, the triangle inequality holds also on the cylinder (the other possible distance from A to C through the cut is chosen only if it makes $d(A, C)$ shorter). Therefore, our ground distance is metric on a cylinder.

To prove that the EMD is metric in the rotation- and scale-invariant cases, we notice that in log-polar space, rotation and scale shifts are reduced to translations (assuming that the ground distance takes care of the cyclic cylinder axis), and we now prove the following stronger theorem:

Denote the translation-invariant EMD between signatures P and Q by

$$\text{EMD}(P, Q, T_{PQ}) = \sum_{i,j} f_{ij} d(p_i, q_j - T_{PQ}) ,$$

where f_{ij} and T_{PQ} are the flow and translation that minimize the sum.

THEOREM 4.1 *If signatures P and Q have equal weight then $EMD(P, Q, T_{PQ})$ is metric.*

Proof: Let $\text{EMD}(P, R, T_{PR})$ and $\text{EMD}(R, Q, T_{RQ})$ be the translation-invariant EMD between signatures P and R, and between signatures R and Q respectively. We need to prove that

$$\text{EMD}(P, Q, T_{PQ}) \leq \text{EMD}(P, R, T_{PR}) + \text{EMD}(R, Q, T_{RQ}) \ .$$

Without loss of generality, we assume that the total sum of the flows is 1. Consider the flow $P \mapsto R \mapsto Q$. The largest unit of weight that moves together from P to R and from R to Q defines a flow which we call b_{ijk} where i, j and k correspond to p_i, r_j and q_k respectively. Clearly $\sum_k b_{ijk} = f_{ij}$ and $\sum_i b_{ijk} = g_{jk}$. We define

$$h_{ik} \triangleq \sum_j b_{ijk}$$

which is a legal flow from i to k since

$$\sum_i h_{ik} = \sum_{i,j} b_{ijk} = \sum_j g_{jk} = w_{r_k} \ ,$$

and

$$\sum_k h_{ik} = \sum_{j,k} b_{ijk} = \sum_j f_{ij} = w_{p_i} \ .$$

Since $\text{EMD}(P, Q, T_{PQ})$ is the minimal flow from P to Q, and h_{ik} is some legal flow from P to Q,

$$\text{EMD}(P, Q, T_{PQ}) \leq \sum_{i,k} h_{ik} d\left(p_i, q_k - (T_{PR} + T_{RQ})\right)$$

$$= \sum_{i,j,k} b_{ijk} d\left(p_i, q_k - (T_{PR} + T_{RQ})\right)$$

$$\leq \sum_{i,j,k} b_{ijk} d(p_i, r_j - T_{PR}) +$$

$$\sum_{i,j,k} b_{ijk} d\left(r_j - T_{PR}, q_k - (T_{PR} + T_{RQ})\right)$$

$$= \sum_{i,j,k} b_{ijk} d(p_i, r_j - T_{PR}) + \sum_{i,j,k} b_{ijk} d(r_j, q_k - T_{RQ})$$

$$= \sum_{i,j} f_{ij} d(p_i, r_j - T_{PR}) + \sum_{j,k} g_{jk} d(r_j, q_k - T_{RQ})$$

$$= \text{EMD}(P, R, T_{PR}) + \text{EMD}(R, Q, T_{RQ}) \ .$$

∎

2.5. Examples

In this section we show examples of the various distances we defined for homogeneous textures, using a Gabor filter bank with five scales and eight orientations.

We use the 16 texture patches shown in Figure 4.4, with the first 13 textures from the VisTex texture collection from the MIT Media Lab[1]. The distance matrices with distances between all pairs of texture patches are shown in tables 4.1-4.3 for the non-invariant, rotation-invariant, and rotation- and scale-invariant distances respectively. The relative distances clearly match human perception. For example, similar textures, such as the pairs {(1), (2)} and {(8),(9)}, result in small distances. Textures that are similar up to rotation, such as the pair {(4), (5)}, are far when no invariance is applied, and close when the rotation-invariant version is used. Similarly, textures that differ in scale, such as the pairs {(10), (11)} and {(12), (13)}, and textures that differ both by scale and orientation such as the pair {(14), (15)}, are close only when the rotation- and scale-invariant version is being used.

	1	2	3	4	5	6	7	8	9	10	11	12	13	14	15	16
1	0.00															
2	0.39	0.00														
3	2.95	2.97	0.00													
4	1.82	1.96	2.35	0.00												
5	1.84	1.75	4.06	1.67	0.00											
6	1.07	1.14	2.08	2.77	2.89	0.00										
7	1.94	1.78	2.55	3.35	3.22	0.99	0.00									
8	1.10	0.92	2.26	2.84	2.65	0.39	1.07	0.00								
9	1.14	0.80	2.23	2.68	2.47	0.52	1.11	0.23	0.00							
10	1.00	0.74	3.07	2.70	2.24	1.11	1.74	0.90	0.89	0.00						
11	2.02	1.71	2.79	3.69	3.17	1.15	1.18	0.91	0.97	1.01	0.00					
12	0.94	1.07	3.38	1.57	1.68	1.91	2.88	1.86	1.76	1.55	2.56	0.00				
13	2.31	2.07	2.38	3.17	3.60	1.39	1.34	1.32	1.27	2.03	1.17	2.29	0.00			
14	3.84	3.86	0.80	3.23	4.92	2.86	2.62	3.05	3.11	3.95	3.17	4.22	2.13	0.00		
15	2.19	2.04	4.66	3.34	2.27	2.57	2.87	2.37	2.31	1.48	2.31	2.55	3.45	5.50	0.00	
16	1.51	1.48	1.96	2.96	3.17	0.76	1.19	0.76	0.82	1.55	0.93	2.32	0.88	2.41	3.08	0.00

Table 4.1.   Distance matrix of the textures in Figure 4.4.

3.   NON-HOMOGENEOUS TEXTURES

Although summarizing the texture content of an image by a single texture descriptor leads to good distance measures between homogeneous textures, most of the information contained in the distribution of the texture features over the image is ignored. The texture content of an image entails a distribution of texture features, even if the image contains only one homogeneous texture. This distribution accounts for four sources of variation in the filter responses:

1 The size of the basic texture element ("texton") is often larger than the support of at least the finest scale Gabor filters. This causes a variation in the filter responses even within textures that a human would perceive as homogeneous in the image. To address this variation, many texture analysis

---

[1]The VisTex texture collection can be obtained from
http://www-white.media.mit.edu/vismod/imagery/VisionTexture/vistex.html

Figure 4.4. 16 homogeneous textures.

methods (for instance, [1, 101, 55]) integrate filter responses over areas that are larger than the largest filter support.

2 Texture regions that a human would perceive as being homogeneous *in the world* can produce inhomogeneous regions *in the image* because of foreshortening and variations in illumination. This spreads the distribution in texture space and increases its variability.

3 Textures exhibit spatial variation even in the world. For instance, most natural textures are regular only in a statistical sense, so filter responses will vary regardless of viewing conditions.

	1	2	3	4	5	6	7	8	9	10	11	12	13	14	15	16
1	0.00															
2	0.23	0.00														
3	2.61	2.38	0.00													
4	1.40	1.46	1.80	0.00												
5	1.60	1.66	2.27	0.54	0.00											
6	1.07	0.93	1.69	2.25	2.45	0.00										
7	1.94	1.78	1.89	2.77	3.00	0.99	0.00									
8	1.08	0.92	1.67	2.27	2.47	0.28	1.07	0.00								
9	0.99	0.80	1.64	2.21	2.41	0.27	1.07	0.18	0.00							
10	0.93	0.74	1.70	1.38	1.62	0.96	1.64	0.90	0.89	0.00						
11	1.93	1.71	1.17	2.21	2.53	0.95	1.13	0.91	0.97	1.01	0.00					
12	0.80	0.94	3.19	1.53	1.54	1.67	2.13	1.82	1.73	1.51	2.56	0.00				
13	2.04	1.88	2.11	3.11	3.33	1.14	0.84	1.10	1.16	1.80	0.96	2.29	0.00			
14	3.50	3.27	0.80	2.68	3.14	2.48	1.97	2.47	2.53	2.58	1.58	4.04	1.86	0.00		
15	2.10	2.04	1.06	1.19	1.60	2.35	2.87	2.37	2.31	1.48	2.31	2.48	3.24	1.56	0.00	
16	1.51	1.44	1.93	2.96	3.17	0.70	1.06	0.69	0.72	1.53	0.90	2.19	0.82	2.41	3.07	0.00

Table 4.2.   Distance matrix of the textures in Figure 4.4 with rotation invariance.

	1	2	3	4	5	6	7	8	9	10	11	12	13	14	15	16
1	0.00															
2	0.23	0.00														
3	1.98	1.91	0.00													
4	1.40	1.46	0.94	0.00												
5	1.39	1.33	0.92	0.54	0.00											
6	0.55	0.54	1.69	1.62	1.50	0.00										
7	1.27	1.22	1.89	1.76	1.83	0.97	0.00									
8	0.53	0.50	1.67	1.62	1.46	0.28	1.07	0.00								
9	0.50	0.48	1.64	1.54	1.43	0.27	1.01	0.18	0.00							
10	0.93	0.74	1.33	0.89	0.90	0.96	1.27	0.90	0.89	0.00						
11	1.18	1.01	1.17	0.97	1.02	0.95	1.13	0.91	0.97	0.35	0.00					
12	0.80	0.94	2.08	1.53	1.54	0.92	0.83	1.04	0.94	1.02	0.96	0.00				
13	0.90	0.96	1.95	1.69	1.63	0.91	0.84	0.83	0.77	1.12	0.96	0.41	0.00			
14	2.44	2.38	0.49	1.34	1.42	2.14	1.97	2.11	2.05	1.91	1.58	2.24	1.86	0.00		
15	2.10	2.04	0.26	1.10	1.06	1.97	2.19	1.96	1.96	1.48	1.39	2.22	2.22	0.45	0.00	
16	0.87	0.84	1.93	1.86	1.79	0.70	1.06	0.69	0.72	0.99	0.90	0.74	0.82	2.41	2.21	0.00

Table 4.3.   Distance matrix of the textures in Figure 4.4 with rotation and scale invariance.

4 Images with multiple textures result in a combination of the distributions of their constituent textures.

3.1.   Texture Signatures

Because of these sources of variation, a single image can produce nearly as many texture vectors as it has pixels. To represent the full distribution of image texture in a compact way, we first find dominant clusters in the $M \cdot L$ dimensional texture space, where M and L are again the number of scales and orientations, respectively, in our texture representation.

We use a K-means clustering algorithm [33], where the number of clusters, K, is predefined. For simple images, such as those of homogeneous textures, a small number of clusters suffices, while complex ones, such as images containing multiple textures, needs a larger number of cluster to reliably represent the distribution in texture space.

While this is a simple and fast algorithm, so that large number of images can be processed quickly, it is conceivable that more sophisticated clustering algorithms that adjust the number of clusters to the image complexity (*e.g.*, the clustering done in [4]) can further improve our texture similarity methods. A variable number of clusters is desirable since the EMD can be applied to signatures of different sizes. The effect of the number of clusters is demonstrated in Figure 4.5 that shows the pixels' assignments to clusters for a natural image using the clustering algorithm with 4, 8, and 16 clusters. Clearly, as the number of clusters increases, more subtle differences are captured. For example, when only four clusters are used, one cluster is used to represent horizontal stripes, and one to represent vertical stripes. Increasing the number of clusters to sixteen, results in separate clusters for different scales of horizontal and vertical stripes. Notice also that variation in the grass texture only starting to get captured with sixteen clusters. As we show in Section 3.4, for natural images we use 32 clusters. In Section 4 we show that an appropriate preprocessing of the images can reduce the number of clusters needed.

The resulting set of cluster centers, together with the fractional cluster weights, is the *texture signature* of the image. An example of a texture signature is shown in Figure 4.6. Part (a) shows an homogeneous texture. Part (b) shows its average texture feature which is used for the homogeneous texture distances defined in Section 2. Part (c) shows the texture signature with four clusters. The cluster representations are shown together with the cluster weights. Part (d) shows the result of assigning every pixel in the image to its cluster.

3.2. Texture-Based Retrieval

We constructed a database of 1792 texture patches by dividing each of the 112 textures from the Brodatz album [9] into 4-by-4 non-overlapping patches, each of size 128-by-128 pixels. All the patches from the Brodatz album are shown in Figures 4.7 and 4.7. The K-means algorithm was used with $K = 8$, resulting in signatures with 8 clusters.

Having defined texture signatures, we can now use the EMD to retrieve images with textures. We use the saturated ground distance defined in Section 4, with the L_1 distance as the underlying dissimilarity. Figure 4.8 shows two examples of retrieving texture patches by using a texture patch as the query. In the first example, the top 16 matches were the patches from the same Brodatz texture. The second example is harder, as the texture used for the query is made of a few sub-textures, and other patches from this Brodatz texture have different amounts of these sub-textures.

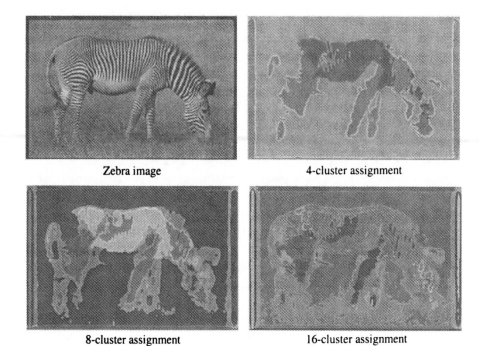

Zebra image   4-cluster assignment

8-cluster assignment   16-cluster assignment

Figure 4.5.   Mapping of image pixels to their nearest texture cluster, differentiated by color. *(This is a color figure)*

3.3.   Partial Matches

An important advantage of the EMD over other measures for texture similarity is its ability to handle images that contain more than one texture without
first segmenting them, as required when using other measures. Using the EMD
for partial matches can find multi-textured images that contain specific textures.
Figure 4.9 shows an example of a partial query. Here we added images with
compositions of textures to our texture database. The query was 20% of the
texture in part (a) and 80% "don't care". The best matches are shown in part
(b) with the 16 patches from the same texture at the beginning followed by all
the compositions that contain some part of the queried texture. We emphasize
again that no segmentation was performed. Figure 4.10 demonstrates a partial
query where the query has more than one texture.

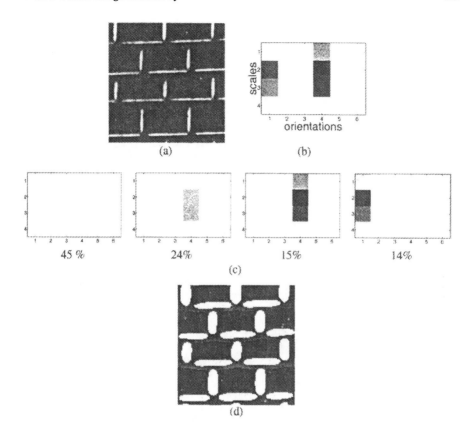

Figure 4.6. (a) Texture patch from the Brodatz album [9]. (b) Average over all texture features. The Gabor filter bank consists of four scales and six orientations. (c) The four clusters in the texture signature. (d) Pixel assignment using the signature's clusters. Different gray levels represent different clusters.

3.4. Retrieving Natural Images

In the next experiment we created a database of 500 grayscale images of animals from the Corel Stock Photo Library[2] with image sizes of 768-by-512 pixels. The K-means algorithm was used with $K = 8$, resulting in signatures with 32 clusters.

Figure 4.11(a) shows an example of a query that used a rectangular patch from an image of a zebra. We asked for images with at least 20% of this texture. The 12 best matches, shown in part (b) and ranked by their similarity to the query,

---

[2]The Corel Stock Photo Library consists of 20,000 images organized into sets of 100 images each. We created our database using the following sets: 123000 (Backyard Wildlife), 134000 (Cheetahs, Leopards & Jaguars), 130000 (African Specialty Animals), 173000 (Alaskan Wildlife), and 66000 (Barnyard Animals).

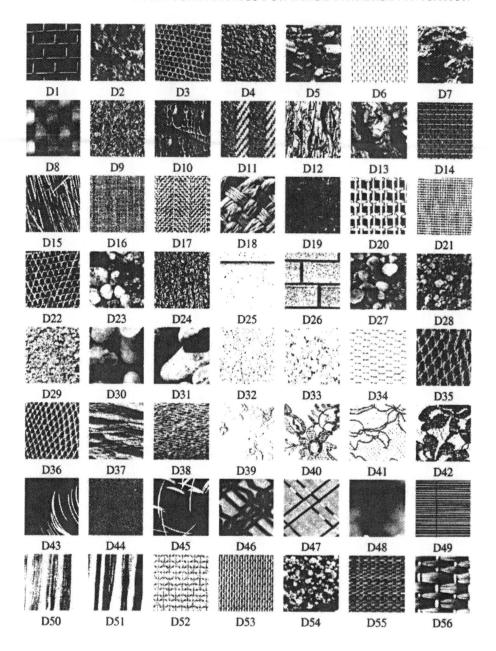

Figure 4.7.   The Brodatz album of textures. Only 1/16 of each texture is shown.

are all images of zebras. In fact, the 16 best matches were all images of zebras, out of a total of 34 images of zebras in the database. The various backgrounds in the retrieved images were ignored by the system because of the EMD's ability to

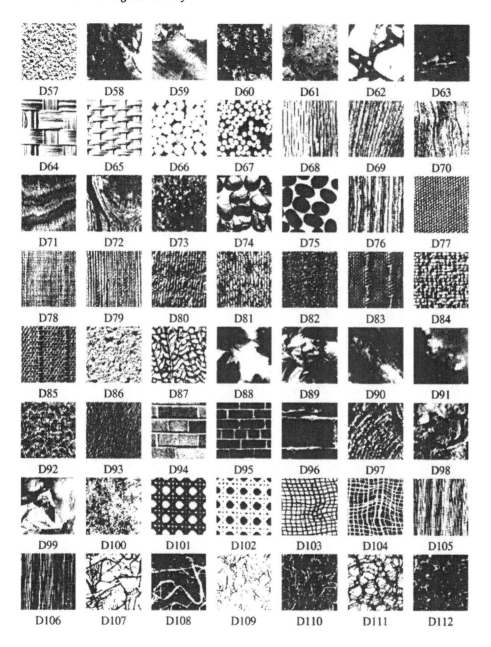

Figure 4.7 (continued). The Brodatz album of textures.

handle partial queries. Notice also that in some of the retrieved images there are a few small zebras, which only when combined together provide a significant amount of "zebra texture." Methods based on segmentation are likely to have problems with such images.

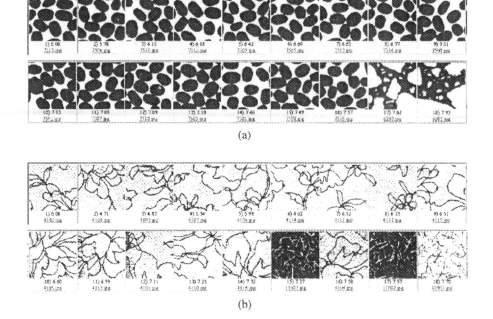

Figure 4.8.   Texture queries. The first image in each part was used as the query. (a) Coffee beans (D75). All 16 patches from the same texture were returned first. (b) Lace (D41). Here, 15 out of the 16 texture patches are in the top 18 matches.

Next we searched for images of cheetahs. The database has 33 images of cheetahs, and 64 more images of leopards and jaguars that have similar texture as cheetahs. Figure 4.12 shows the query and the best matches. The first eight images are indeed cheetahs. The other four matches are images of leopards and jaguars.

To check if our method can distinguish between the different families of wild cats, we looked for images of jaguars. Figure 4.13 shows the query results. From the best twelve matches, eleven are jaguars and leopards, which are almost indistinguishable. Only the sixth match was an image of a cheetah.

4.   IMAGE PREPROCESSING

The texture features introduced in Section 1 describe the local appearance of small image neighborhoods. The size of these neighborhoods is equal to that of the supports of the coarser resolution filters employed. On the one hand, these supports are large enough that they can straddle boundaries between different textures. In this case, they do not describe "pure" textures, and they can convey information that is hard to interpret and can be misleading. On the other hand,

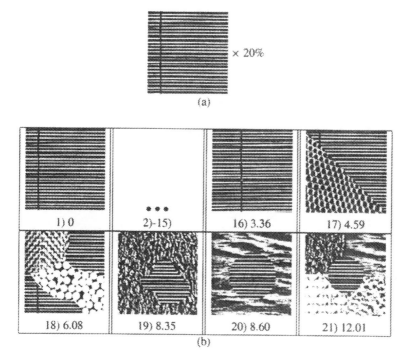

Figure 4.9. Partial texture query. The query was 20% of the texture in part (a) and 80% "don't care". (b) The 21 best matches: 16 patches from the same texture (only the first and last ones are shown), followed by all the compositions that contain some part of the queried texture.

the supports are often too small for them to "see" enough of a texture to yield a reliable description. In fact, the basic period of repetition of the underlying texture may be comparable to the filter support sizes, so that adjacent filters see somewhat different parts of the texture, and the description vectors differ somewhat. Also, textures exhibit variations in the world, as well as variations caused by nonuniform viewing parameters.

Before using texture features to compute texture signatures, it is therefore preferable to sift and summarize the information that they convey. Specifically, it is desirable to eliminate vectors that describe mixtures of textures, and to average away some of the variations between adjacent descriptors of similar image patches.

We propose to let descriptors of similar and adjacent texture neighborhoods *coalesce* into tight clusters. In this approach, texture features are first computed for the image. If two adjacent features describe similar textures, a smoothing process brings them even closer to each other. If the texture descriptors are very different, however, then they are modified so that their distance is left unchanged or is even increased.

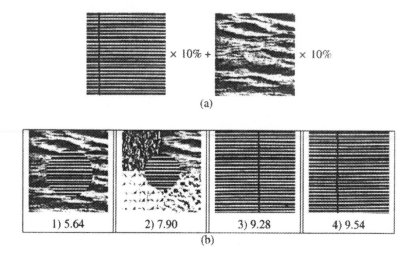

(a)

1) 5.64   2) 7.90   3) 9.28   4) 9.54

(b)

Figure 4.10.   Another partial query. The query now contains 10% of each of the two patches in part (a) and 80% "don't care." (b) The two best matches are the two compositions that contain the textures in the query, followed by the patches that contain only one of the queried textures.

In other words, we propose to do edge-preserving smoothing, but in the space of texture vectors rather than image intensities. This requires generalizing the definition of gradient to a vector function, which we do by introducing a measure of *texture contrast*. Thanks to this definition, we can run the equivalent of anisotropic diffusion processes [68] to achieve edge-preserving smoothing. We use the texture contrast also to define *significant regions* as regions where contrast is low after a few iterations of the smoothing process. Only texture features that are in a significant region are used for the computation of the texture signature.

4.1.   Texture Contrast

For texture contrast, we employ a notion of "generalized gradient" from differential geometry [48] that has been used for color images in the past few years [21, 15, 86]. The rest of this section discusses this notion as it applies to texture.

Assume that we have a mapping $\Phi : S \subseteq \Re^n \to \Re^m$. Let ϕ_i denote the ith component of Φ. If Φ is a texture vector space, for example, then $\phi_i, 1 \leq i \leq m$ are the responses to the Gabor filters, and $n = 2$ is the image dimensionality. If Φ admits a Taylor expansion we can write

$$\Phi(\mathbf{x} + \Delta\mathbf{x}) = \Phi(\mathbf{x}) + \Phi'(\mathbf{x})\Delta\mathbf{x} + \|\Delta\mathbf{x}\|e(\mathbf{x}, \Delta\mathbf{x}) \ .$$

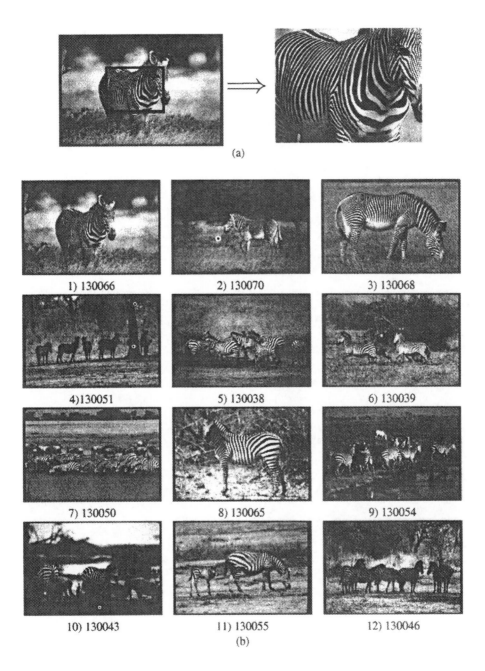

Figure 4.11. Looking for zebras. (a) An image of a zebra and a block of zebra stripes extracted from it. (b) The best matches to a query asking for images with at least 10% of the texture in (a). The numbers in the thumbnail captions are indices into Corel CDs.

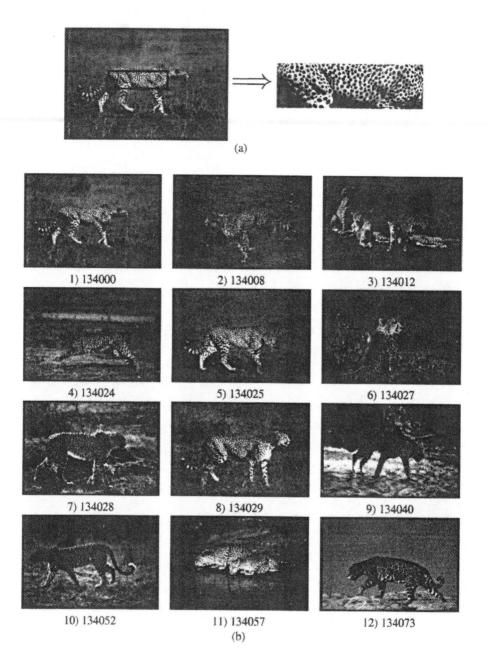

(a)

1) 134000   2) 134008   3) 134012

4) 134024   5) 134025   6) 134027

7) 134028   8) 134029   9) 134040

10) 134052   11) 134057   12) 134073

(b)

Figure 4.12. Looking for cheetahs. (a) The query. (b) The best matches with at least 10% of the query texture. The last four images are leopards and jaguars which have similar texture as cheetahs. However, cheetahs come first.

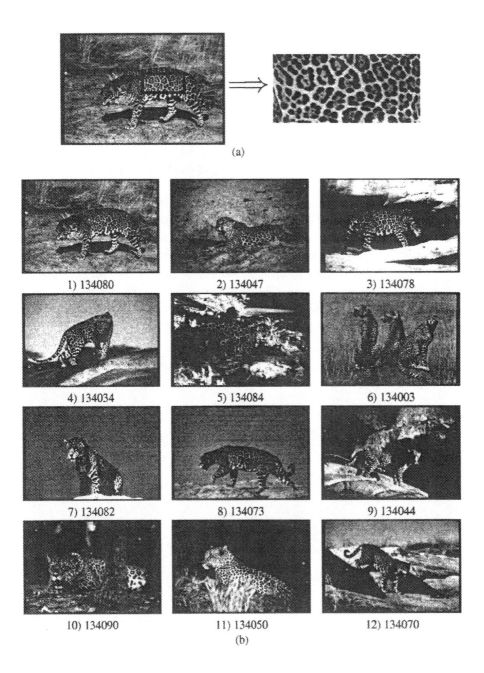

(a)

1) 134080 2) 134047 3) 134078

4) 134034 5) 134084 6) 134003

7) 134082 8) 134073 9) 134044

10) 134090 11) 134050 12) 134070

(b)

Figure 4.13. Looking for leopards and jaguars. (a) The query. (b) The best matches with at least 10% of the query texture. All but the sixth image are leopards and jaguars. The sixth image is of cheetahs.

where $\|e(\mathbf{x}, \Delta\mathbf{x})\| \to 0$ as $\Delta\mathbf{x} \to 0$ and $\Phi'(\mathbf{x})$ is the $m \times n$ Jacobian matrix of Φ:

$$\Phi'(\mathbf{x}) = J(\mathbf{x}) = \begin{bmatrix} \frac{\partial \phi_1}{\partial x_1} & \cdots & \frac{\partial \phi_1}{\partial x_n} \\ \vdots & & \vdots \\ \frac{\partial \phi_m}{\partial x_1} & \cdots & \frac{\partial \phi_m}{\partial x_n} \end{bmatrix} .$$

If one starts at point $\mathbf{x}$ and moves by a small step $\Delta\mathbf{x}$, the distance traveled in the attribute domain is approximately

$$d \cong \|\Phi'(\mathbf{x})\Delta\mathbf{x}\| = \sqrt{\Delta\mathbf{x}^T J^T J \Delta\mathbf{x}} .$$

The step direction which maximizes d is the eigenvector of $J^T J$ corresponding to its largest eigenvalue. The square root of the largest eigenvalue, or, equivalently, the largest singular value of Φ', corresponds to the gradient magnitude, and the corresponding eigenvector is the gradient direction.

We can give a closed form solution for the eigenvalues in the case that $n = 2$, as it is for images. In this case, Φ is the mapping from the two-dimensional image plane to the M-by-L dimensional texture space, where M and L are the number of scales and orientations used. The differential of Φ is

$$d\Phi = \sum_{i=1}^{2} \frac{\partial \Phi}{\partial x_i} dx_i .$$

and so

$$\|d\Phi\|^2 = \sum_{i=1}^{2} \sum_{j=1}^{2} \frac{\partial \Phi}{\partial x_i} \cdot \frac{\partial \Phi}{\partial x_j} dx_i dx_j .$$

Using the notation from Riemannian geometry [48], we have

$$\begin{aligned} \|d\Phi\|^2 &= \sum_{i=1}^{2} \sum_{j=1}^{2} g_{ij} dx_i dx_j \\ &= \begin{bmatrix} dx_1 & dx_2 \end{bmatrix} \begin{bmatrix} g_{11} & g_{12} \\ g_{21} & g_{22} \end{bmatrix} \begin{bmatrix} dx_1 \\ dx_2 \end{bmatrix} . \end{aligned}$$

where

$$g_{ij} \equiv \sum_{k=1}^{m} \frac{\partial \phi_k}{\partial x_i} \frac{\partial \phi_k}{\partial x_j} .$$

and $g_{12} = g_{21}$. Now let

$$\lambda_{\pm} = \frac{1}{2} \left(g_{11} + g_{22} \pm \sqrt{(g_{11} - g_{22})^2 + 4g_{12}^2} \right) .$$

be the two eigenvalues in the matrix $G = \begin{bmatrix} g_{11} & g_{12} \\ g_{21} & g_{22} \end{bmatrix}$. Since G is real and symmetric, its eigenvalues are real.

We choose λ_+ as the generalized gradient magnitude since it corresponds to the direction of maximum change. We can verify that this magnitude reduces to the ordinary gradient norm in the case of a grayscale image ($m = 1$):

$$
\begin{aligned}
\lambda_+ &= \frac{1}{2} \left(\Phi_x^2 + \Phi_y^2 + \sqrt{(\Phi_x^2 - \Phi_y^2)^2 + 4\Phi_x^2\Phi_y^2} \right) \\
&= \frac{1}{2} \left(\Phi_x^2 + \Phi_y^2 + (\Phi_x^2 + \Phi_y^2) \right) \\
&= \Phi_x^2 + \Phi_y^2 = \|\nabla\Phi\|^2
\end{aligned}
$$

where the subscripts x and y denote differentiation.

While the application of this definition to color is relatively straight forward, matters are more complicated for texture. Because of the multiscale nature of texture in general, and of our texture descriptors in particular, different filters have different supports. Computing derivatives at different scales requires operators of appropriate magnitudes and spatial supports in order to yield properly scaled components. For instance, if differentiation is performed by convolution with the derivatives of a Gaussian, the standard deviation of each Gaussian must be proportional to scale, and the magnitudes must be properly normalized. Figure 4.14(b) shows the texture contrast of the texture mosaic from Figure 4.14(a). The texture boundaries are characterized by high contrast.

Figure 4.14(b) shows some areas with relatively high texture contrast even inside regions that appear to be of the same texture. The sources of these contrast areas have been discussed at the beginning of Section 4, where it was pointed out that small, low-contrast areas should be smoothed away, while extended ridges of the contrast function should be maintained. The measure of texture contrast introduced above allows extending anisotropic diffusion [68] or edge-preserving smoothing [84] techniques to texture descriptors.

4.2. Edge-Preserving Smoothing

Here, we present an efficient implementation of an edge-preserving smoothing algorithm, which is based on repeatedly convolving each of the spectral subbands of the image with a separable binomial filter weighted by a measure of the texture contrast. Efficiency considerations are particularly important in an application like image retrieval, where large numbers of images must be processed when they are entered into the database. After describing the algorithm, we explore its relation with the method described in [68].

First, the weighting function $g(C)$ is introduced, where C is the texture contrast defined in Section 4.1. The function $g(C)$ should be high where the texture is uniform and low where the texture is "edgy" and can be any non-negative monotonically decreasing function with $g(0) = 1$. We chose to use

$$g(C) = e^{-\left(\frac{|C|}{k}\right)^2} \tag{4.5}$$

where k controls the decay rate of $g(\cdot)$ and, as we will see later, determines which of the edges is preserved.

Smoothing is performed by convolving the spectral subbands of the image with the binomial filter

$$\mathbf{B} = B^T B , \qquad B = \begin{bmatrix} 1 & 2 & 1 \end{bmatrix} ,$$

after it is weighted by $g(C)$. We chose this filter because it is separable and can be applied efficiently. When applied repeatedly, the binomial filter quickly gives an approximation of a Gaussian. For brevity, we introduce the following notation:

$$a *_x b = \sum_{i=-1}^{1} a(i)b(x + i, y) ,$$

$$a *_y b = \sum_{j=-1}^{1} a(j)b(x, y + j) ,$$

$$c * b = \sum_{i=-1}^{1} \sum_{j=-1}^{1} c(i, j)b(x + i, y + j) .$$

A single iteration of the smoothing procedure is computed at time t as follows:

1 Compute the texture contrast $C^{(t)}$.

2 Set $G^{(t)} = g(C^{(t)})$.

3 For $0 \le m < M$ and $0 \le l < L$ do:

   (a) Let $I^{(t)}$ be the m-th scale and the l-th orientation subband.

   (b) Compute:

$$
\begin{aligned}
K^{(t)} &= B *_x G^{(t)} , \\
J^{(t)} &= \frac{B *_x (G^{(t)} I^{(t)})}{K^{(t)}} , \\
P^{(t)} &= B *_y K^{(t)} , \\
I^{(t+1)} &= \frac{B *_y (K^{(t)} J^{(t)})}{P^{(t)}} .
\end{aligned}
$$

(This computation can be simplified. For clarity, we show it in the above form).
It is easy to see that

$$I^{(t+1)} = \frac{\mathbf{B} * (G^{(t)} I^{(t)})}{\mathbf{B} * G^{(t)}} .$$

Without the $\mathbf{B}(i, j)$ factor, this is similar to the adaptive smoothing proposed in
[84], but implemented, in addition, in a separable fashion for greater efficiency.
In [84], this iterative process is proven to be stable and to preserve edges. A
similar proof can be applied to our case with straight forward modifications.
That edges are preserved is proved by showing that when the contrast is large
enough ($> k$), it increases as the iterations progress, thereby sharpening the
edge. When the contrast is small ($< k$), the contrast decreases and the edge is
smoothed. This implies that k is equivalent to a contrast threshold.

In the following, we show that the edge-preserving smoothing iteration is
equivalent to an iteration of the anisotropic diffusion proposed by [68]. Define

$$c^{(t)} = \frac{G^{(t)}}{\mathbf{B} * G^{(t)}} .$$

Then we have

$$I^{(t+1)} = \mathbf{B} * (c^{(t)} I^{(t)})$$

and by using the fact that

$$\mathbf{B} * c^{(t)} = 1 ,$$

we can write

$$I^{(t+1)} - I^{(t)} = \mathbf{B} * [c^{(t)} (I^{(t)} - I^{(t)} \delta)] , \qquad (4.6)$$

where $\delta(x, y) = 1$ if $x = y = 0$ and 0 otherwise. Equation (4.6) is a discretization of the anisotropic diffusion equation

$$I_t = \nabla \cdot (c(x, y, t) \nabla I) .$$

Instead of using a 4-neighbor discretization of the Laplacian as done in [68],
we use a better, 8-neighbor discretization [43]:

$$\frac{1}{4} \begin{bmatrix} 1 & 2 & 1 \\ 2 & -12 & 2 \\ 1 & 2 & 1 \end{bmatrix} .$$

Figure 4.14(c) shows the texture contrast after 20 iterations. To visualize the texture vectors we project them onto the plane spanned by the two
most significant principal components of all texture vectors in the image. Figures 4.14(d), (e) show the projections of the texture descriptors before and after
the edge-preserving smoothing. Only descriptors from *significant regions* are
shown in the latter. A region is significant if the contrast after smoothing is

everywhere smaller than the contrast threshold k in Equation (4.5). We can see that the descriptors of the four textures form clear, distinct clusters even in this two-dimensional projection. Notice the sparse trail of points that connects the two left-most and the two top-most clusters in Figure 4.14(e). These points come from a "leakage" in the boundary between two textures. This implies that we are limited in the amount of smoothing we can do before different textures start to mix. This is also a situation where most segmentation algorithms would have to make a classification decision, even if the two textures involved are similar to each other.

Figure 4.15(b) shows the significant regions of the image in Figure 4.15(a) after ten iterations. Texture features that belong to insignificant regions are likely to be on the boundary of more than one texture, and should be ignored when computing the texture signature, as they are not good representatives for any of the textures.

5.   SUMMARY

In this chapter we defined families of texture metric distances, based on the EMD, and showed by examples that they match perceptual similarity well. The texture features are based on carefully designed Gabor filters. For homogeneous texture patches it was enough to summarize the texture content by one descriptor, although using the full distribution of texture features captures the texture content better. We also presented rotation- and scale-invariant versions of the texture metric for homogeneous textures. Non-homogeneous textures, such as those found in natural images, required using the full distribution of the texture features, and introduced problems due to the inherent spatial extent of texture, such as the blending of adjacent textures that result in meaningless texture features. We were able to filter out these features by defining a measure of texture contrast. Using the texture contrast also led to a smoothing algorithm that coalesced texture features that belonged to the same texture, without blending features from neighboring textures.

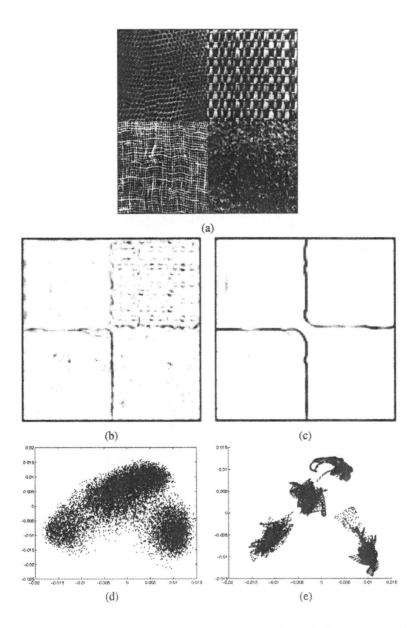

Figure 4.14. (a) Mosaic of four textures from the Brodatz album. (b) Texture contrast before smoothing (dark values correspond to large contrast) and (c) after 20 iterations. (d) Projection of the texture descriptors onto the plane of their two most significant principal components before smoothing and (e) after 20 iterations.

(a)   (b)

Figure 4.15.   (a) A picture of a lighthouse. (b) Nonignificant regions (blackened out) after ten iterations of smoothing.

Chapter 5

COMPARING DISSIMILARITY MEASURES

Doubt, the essential preliminary of all improvement and discovery, must accompany the stages of man's onward progress. The faculty of doubting and questioning, without which those of comparison and judgment would be useless, is itself a divine prerogative of the reason.

—Albert Pike, 1809–1891

In Chapter 1 we surveyed distribution-based dissimilarity measures and discussed their properties, and in Chapter 2 we presented the EMD, which we hypothesized to be a good choice for image retrieval applications. In Chapters 3 and 4 we showed how these measures can be instantiated to compare color and texture distributions In this chapter we compare the results of the different dissimilarity measures when used for color- and texture-based image retrieval. The main difficulty in such a comparison is establishing ground truth that will help to determine if a returned image is relevant or not. To this end, we create databases where the ground truth is known, and use them to conduct experiments that evaluate the performance of the dissimilarity measures.

1. BENCHMARK METHODOLOGY

Any systematic comparison of dissimilarity measures should conform at least to the following guidelines:

- A meaningful *quality measure* must be defined. Different tasks often entail different quality measures. For image retrieval, performance is usually measured by *precision*, which is the number of relevant images retrieved relative to the total number of retrieved images, and *recall*, which is the number of relevant images retrieved, relative to the total number of relevant images in the database. Plotting the precision vs. the number of retrieved images, for all dissimilarity measures, is a good way to compare between them. However, in order to decide if a particular measure is better for a specific task, the precision vs. recall plot might be of better use as the relative importance of good recall vs. good precision differs according to the task at hand. We therefore show both plots for most of our experiments.

- Performance comparisons should account for the variety of *parameters* that can affect the behavior of each measure. These parameters include, in our case, the number of bins in a histogram; the shape and definition of the bins; and for texture, the number of filters used to represent the texture. A fair comparison in the face of this variability can be achieved by giving every measure the best possible chance to perform well. In our experiments we vary these parameters and report their effect on the different dissimilarity measures.

- Processing steps that affect performance independently ought to be evaluated separately in order to both sharpen insight and reduce complexity. For instance, the effect of different histogram sizes can be understood separately from those of different dissimilarity measures.

- *Ground truth* should be available. This is a set of data for which the correct solution for a particular problem is known. Collecting ground truth is arguably the hardest problem in benchmarking, because the "correct solution" ought to be uncontroversial, and the ground-truth data set should be large enough to allow a statistically significant performance evaluation.

2.   EXPERIMENTS

In this section we describe experiments that compare and evaluate the different dissimilarity measures, discussed in this book, for color images and for homogeneous textures.

2.1.   Color

We performed two experiments, each using a different database of color distributions. The first database was constructed such that multiple sets of color distributions were taken from each of a set of color images. The ground truth was defined such that two distributions belong to the same class if they originate from the same image. The second database was the 20,000-image database that we used in Chapters 3 and 4. The ground truth was now established manually by selecting sets of similar images *a priori*, both in the semantic meaning and in their color contents.

In our first experiment, we randomly chose 94 color images[1] from our 20,000 color image database. From each image we created disjoint sets of randomly sampled pixels and considered these sets as belonging to the same class. While for large sets of pixels within a class the color distributions are likely to be very similar, for small sets the variations are larger, mimicking the situation in image retrieval where images of *moderate* similarity to the query have to be identified. To get moderate similarities, we used sets of 8 pixels, and obtained for each image 16 disjoint sets of random samples, resulting in a ground truth data set of 1504 color distributions with 94 different classes, one class per image.

We represented the color distributions in the database by histograms which were adapted to optimally represent the combined distribution over the entire database. To construct the histograms, we ran a K-means clustering algorithm [33] on the combined distribution, resulting in optimal clusters which were used as the histogram prototypes. Each pixel in every color distribution in this database was then assigned to a bin by finding its closest prototype. We repeated this process to compute histograms with 32, 128, and 512 bins, which correspond to coarse, intermediate and fine binning, respectively. For the EMD and the Quadratic Form that require a ground distance, we use the saturated ground distance described in Section 4.

In addition to computing histograms, we also computed signatures for all color distributions. We used the same K-means algorithm as used for histograms, but instead of computing the optimal prototypes of the combined distribution over the entire database, we computed the optimal prototypes for every distribution individually, asking for the best eight prototypes. This resulted in signatures, all with eight clusters.

Once the histograms and signatures were computed for all the distributions in the database, we used the different dissimilarity measures to retrieve and rank the most similar histograms in the database, where each of the 1504 color distributions was used, in turn, as the query. The average of the results of these 1504 queries, for all histogram sizes, was used to evaluate the performance of the dissimilarity measures. Since the EMD is the only method that takes advantage of signatures, we applied the EMD on both histograms and signatures. Figures B.1, B.2, and B.3 in Appendix B show the retrieval performance of the different dissimilarity measures when using histograms with 32, 128, and 512 bins, respectively. We added to each plot the results of the EMD applied to the signatures (with only 8 clusters). The top part of each figure shows the number of relevant distributions in the retrieved set, as a function of the total number of retrieved distributions. The bottom part of the figures shows precision vs. recall.

---

[1] We used the same number of images (94) as in the texture case (see Section 2.2), so that we can compare results from both cases.

Analyzing the results leads to the following conclusions:

- Bin-by-bin dissimilarity measures improve by increasing the number of bins up to a point, after which their performance degrades. Smaller number of bins results in bigger bins. When the bins are too big, the discrimination power of the histograms decreases. When the bins are too small, similar pixels might fall into different bins and will be considered as completely dissimilar. This problem does not exist for cross-bin dissimilarity measures where the similarity of the pixels is determined by the ground distance, small in our case, between the appropriate bins in the feature space.

- For our case of moderate similarities, cross-bin dissimilarity measures perform better than bin-by-bin measures. The EMD that was applied to signatures performed best in all cases, except for the fine, 512-bin histograms case, where it performed slightly better using the histograms. However, for an image retrieval system, using the EMD on 512-bin histograms is too slow. While the EMD on signatures performs best for moderate similarities, once the assumption of moderate similarities is violated and the dissimilarities between objects in the same class get smaller with respect to the dissimilarities between objects from different classes, other dissimilarity measures, such as the χ^2 statistics and Jeffrey divergence, might perform better because they were designed to compare close distributions.

- Although, from an information-theoretic point of view, the signatures in our experiments carry less information than the histograms and can be encoded using fewer bits, they usually lead to better results.

- χ^2 statistics and Jeffrey divergence give almost identical results.

- Among the Minkowski-form distances that we tried, the L_1 distance gave, in general, better results than the L_2 distance, which was better than L_∞. Notice that we measure distances in the histogram space and not in the color space where the L_2 distance, by construction, matches the perceptual similarity between colors.

Representing the distributions by signatures leads, in the case of moderate similarities, to better results than representing them by histograms. However, when distributions from the same class are very similar (in contrast to being moderately similar), adaptive, global histograms perform better than signatures, because features with minor dissimilarities are likely to be assigned to the same bins. This is demonstrated in Figure B.4 where sets of 32 random pixels were taken from the images (instead of 8 pixels as in the case of moderate similarities). Now sets of pixels that originate from the same image are very similar, and almost every measure performs well. The EMD applied to signatures performs slightly worse than most other measures, while the EMD applied to histograms still performs slightly better than the rest of the measures.

The second experiment was conducted on our full 20,000-image database of color images. Our goal in this experiment was to compare different dissimilarity measures using images that are perceived as having similar color content. To do so, we looked for sets of images that have high correlation between their semantic meaning and their color distribution. For the first set, we identified all the images of red cars in the database (75 images) and marked them as relevant. For the second set, we similarly identified images of brown horses in green fields (157 images).

Unlike the first experiment where adaptive histograms were used, here we use histograms with regular binning, as they are commonly used in image retrieval systems and do not require prior knowledge of the database content. We run this experiment twice: once on color histograms with coarse binning, and once with fine binning. For the coarse binning, we divided the CIELab color space into fixed-size cubes of length 25. This quantized the color space into 4 bins in the L channel and 8 bins in both the a and the b channels, for a total of 256 bins. However, most of these bins are always empty due to the fact that valid RGB colors can map only to a subset of this CIELab space. In fact, only 130 bins can have non-zero values, so the histograms have 130 bins. After removing bins with insignificant weights (less than 0.1%), the average histogram has 15.3 non-empty bins. Notice that the amount of information contained in the signatures (8.8 clusters on average) is comparable to that contained in the histograms. For the fine binning, we divided the CIELab color space into fixed-size cubes of length 12.5. This resulted in 2048 bins, of which only 719 can be non-empty. Over our 20,000-image database the average fine histogram has 39 non-empty bins.

In this experiment, instead of using the saturated ground distance (Equation (2.4)) for the EMD and the Quadratic Form, as in the first experiment, we use the Euclidean distance in CIELab as the ground distance. In addition to being faster to compute, we found that for real images, using the Euclidean ground distance for color-based image dissimilarity leads to better recall than when using the saturated ground distance. Being induced by a norm, the Euclidean ground distance also allows the use of the lower bound described in Section 3.2, which significantly reduces the computation time.

From the set of 75 red car images we chose the 10 images shown at the top of Figure 5.2. In these ten images the red car had green and gray in the background, was relatively big (so red is a predominant color), and not obscured by the background. We performed ten queries using these ten images and averaged the number of relevant images for the different dissimilarity measures as a function of the number of images. An example of such a query is shown in Figure 5.1. The color content of the leftmost image of a red car was used as the query, and the eight images with the most similar color contents were returned and displayed in order of increasing distance for different histogram dissimilarity measures. We repeated this process for the set of the 157 horse images, where again, ten "good" images of horses (Figure 5.3, top) were used for the queries.

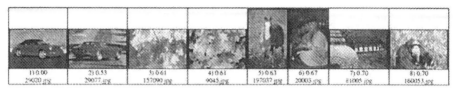

L_1 distance

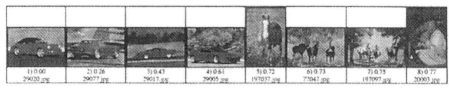

Jeffrey divergence

χ^2 statistics

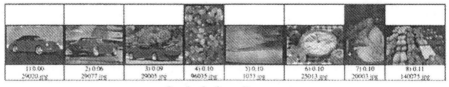

Quadratic-form distance

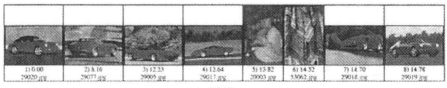

EMD

Figure 5.1. The eight closest images for each of the red car images in the first column. The queries were processed by a color-based image retrieval system using different histogram dissimilarity measures. *(This is a color figure)*

The results of these experiments are shown in Figures 5.2 and 5.3. For the cars, the average number of relevant images for the different dissimilarity measures is shown in Figure 5.2 as a function of the number of images retrieved for coarse (middle) and fine (bottom) histograms respectively. The EMD with histograms outperformed the other histogram-based methods, and the EMD with signatures performed best. The red car images have only partially similar color distributions because only part of an image matches part of the others. The colors of the cars are very similar in all the relevant images while the colors of the backgrounds have more variation. Although other images that do not have cars in them might match the color contents of the query images better, we still expect some of the cars to be retrieved when a large number of images is returned by the system.

For the horses, both the colors of the objects and the colors of the backgrounds are similar for all the relevant images. Figure 5.3 shows the results for coarse and fine histograms. Again, the EMD with signatures performed best. Both the Jeffrey divergence and the χ^2 statistics outperformed the EMD with coarse histograms but not with fine histograms. This can be explained as before by the fact that, for coarser histograms, the ground distance is computed between more distant bin centers, and therefore becomes less meaningful. We recall that only small Euclidean distances in CIELab space are perceptually meaningful. On the other hand, as we saw before, bin-by-bin distances break down as the histograms get finer, because similar features are split among different bins.

2.2. Texture

In order to compare different dissimilarity measures for homogeneous textures, we used the Brodatz album [9]. Patches from the Brodatz album are shown in Figures 4.7 and 4.7. To construct an image database with a known ground truth, each image is considered as a single, separate class. This is a questionable assumption in a few cases, so we *a priori* selected 94 Brodatz textures by visual inspection. We excluded the textures D25, D30-D31, D39-D45, D48, D59, D61, D88-D89, D91, D94, D97 due to the following two problems: the characterizing scales are too coarse to be captured by our filters, or the texture appearance changes over the texture. In these two cases, small blocks from the same textures will lack similarity.

For each texture, we computed the Gabor texture features described in Section 3 and took 16 non-overlapping patches of 8-by-8 pixels with the corresponding texture features. We chose small patches so that patches from the same texture will not be too similar and moderate similarity can be assumed. Similarly to the color experiment, the database contains 1504 texture patches with 94 different classes, each with 16 samples. We used each of the patches in the database as a query, and averaged the results over all the patches. The retrieval performance of the different dissimilarity measures are shown in Figures B.5-

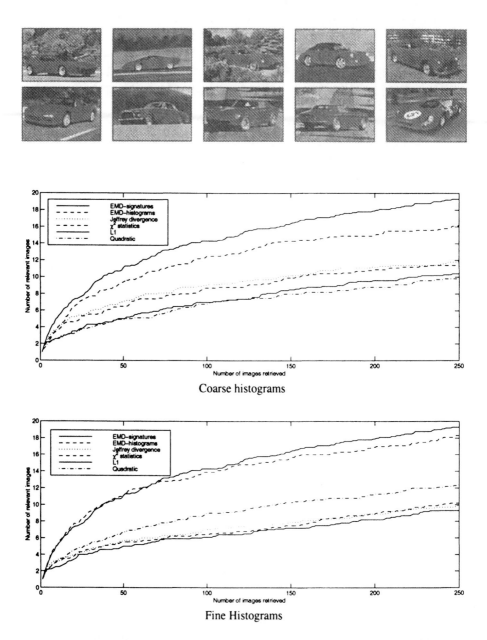

Coarse histograms

Fine Histograms

Figure 5.2. Ten images of red cars (top) and the average number of relevant images, for the different dissimilarity measures, that were returned by using the ten images as the queries with coarse histograms (middle) and fine histograms (bottom). The results obtained using signatures is also shown for reference. *(This is a color figure)*

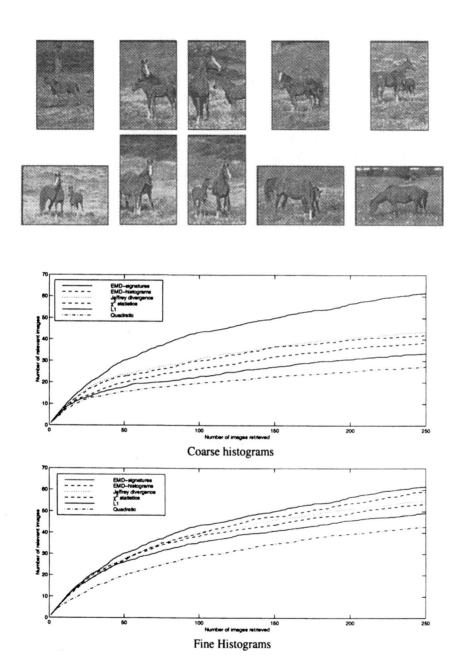

Figure 5.3. Ten images of horses (top) and the average number of relevant images, for the different dissimilarity measures, that were returned by using ten images as the queries with coarse (middle) and fine histograms (bottom). The results obtained by using signatures is also shown for reference. *(This is a color figure)*

B.13 in Appendix B for all combinations of filter bank size and histogram coarseness. For the filter banks we used 12 filters (3 scales/4 orientations), 24 filters (4/6), or 40 filters (5/8); the histograms had 32, 128, or 512 bins. Again, for comparison, we added to all the plots the results of the EMD applied to signatures (with only 8 clusters).

We confirm that our conclusions from the color experiments hold also for texture, and in addition:

- For the smallest filter bank size, the χ^2 statistics and the Jeffrey divergence give the best results. However, increasing the number of filters significantly improves the EMD results, which for the largest filter bank size gives the best results overall. For the EMD, similarly to increasing the number of bins, more filters implies a better representation of the texture space and therefore better results. The performance of the bin-by-bin dissimilarity measures for texture is similar to that for color: as the number of bins increases, it improves up to a point and then degrades. For texture, however, the notion of bin size includes also the number of filters in the filter bank. Since increasing the number of filters is equivalent to sampling the space of orientations and scales more finely, a larger number of filters improves the discrimination power up to a point after which the performance degrades. The optimal points depends on the number of bins used. As can be seen from the graphs, increasing the number of filters improves the results only for the coarse histograms (32 bins) while for fine histograms (512), the performance degrades.

- Examining the precision vs. recall graphs reveals that applying the EMD to signatures gives better results when good precision is wanted, while for good recall, it is better to use the EMD on histograms. Signatures are, by definition, more precise than histograms, so it is natural to expect higher precision in our queries. On the other hand, globally adapted histograms coalesce to the same bins features with only moderate similarity, bringing closer images that have overall moderate similarities between them.

3. SUMMARY

A quantitative performance evaluation was presented for distribution-based image dissimilarity measures. We emphasized image retrieval applications where the ability to compare moderate similarities is of the greatest importance. The EMD performed very well in most cases, and exhibited improvements when more data was available in the form of finer histograms and, for texture, larger filter banks. For the other dissimilarity measures, increasing the amount of data often decreased the performance as a result of inherent binning problems and the curse of dimensionality. Nevertheless, the choice of a dissimilarity measure should be made based on the application at hand and the available data.

Chapter 6

VISUALIZATION

One picture is worth a thousand words.

—Fred R. Barnard

If a picture is worth a thousand words, a picture of an image database is worth a whole book.

—Carlo Tomasi

The images that best match a query should be displayed to the user in a useful way, that shows immediately if there are any relevant images in the returned set. Traditional image retrieval systems display the returned images as a list, sorted by dissimilarity to the query. At this point, the user can examine the images one at a time, and decide the next action. While this might suffice if the correct images are in the returned list, this is usually not the case, even when the user has as exact image in mind. In many cases, the user has only a vague idea of what he is looking for. In this cases, the display should convey information as to what kind of images were returned by the query, what (if anything) went wrong, how to refine the query in order to get closer to the desired images, and whether or not to terminate the search.

There are two shortcomings to the traditional approach. First, when images are ranked by similarity to the query, related images can appear at separate places in the list. Often the user would like to have a global view of the returned images in a way that reflects the relations among the images in the returned set. Second, browsing and navigating in a large database is disorienting unless the user can form a mental picture of the entire database. Only having an idea of the surroundings can offer an indication of where to go next. The wider the horizon, the more secure navigation will be. The small pieces of the database, shown by the traditional methods one at a time, make it harder for the user to form such a picture.

In contrast, we propose a new display technique, which is the main point of this chapter, by which the returned images are displayed not only in order of increasing dissimilarity from the query but also according to their mutual dissimilarities, so that similar images are grouped together rather than being scattered along the entire returned list of images. With such a view, the user can see the relations between the images, better understand how the query performed, and express successive queries more naturally. In brief, the user of the system would benefit from a more coherent view of the query results. In addition, many more images can be returned and displayed with such a view without overloading the user's attention. he user can see larger portions of the database at a glance and form a global mental model of what is in it.

How can such a display be created? If the dissimilarities between all pairs of images in the set of returned images are used, the results of a query can be embedded in a two- or three-dimensional space using *multi-dimensional scaling* techniques, by which image thumbnails are placed on the screen so that screen distances reflect as closely as possible the dissimilarities between the images. If the computed dissimilarities agree with perception, and if the resulting embedding preserves these dissimilarities reasonably well, than the resulting display should be perceptually intuitive.

While traditional displays list images in order of similarity to the query, thereby representing only n dissimilarities if n images are returned, our display conveys information about all $\binom{n}{2}$ dissimilarities between images. Our geometric embeddings allow the user to perceive the dominant axes of variation in the displayed image group. Thus, the embeddings are *adaptive*, in the sense that they use the screen's real estate to emphasize whatever happen to be the main differences and similarities among the particular images at hand.

1.   MULTI-DIMENSIONAL SCALING

Given a set of n objects together with the dissimilarities δ_{ij} between them, the multi-dimensional scaling (MDS) technique [100, 88, 49] computes a configuration of points $\{p_i\}$ in a low-dimensional Euclidean space $\mathbf{R}^d$, (we use $d = 2$ or $d = 3$) so that the Euclidean distances $d_{ij} = \|p_i - p_j\|$ between the points in $\mathbf{R}^d$ match the original dissimilarities δ_{ij} between the corresponding objects as well as possible. Kruskal's formulation of this problem [49] requires minimizing the following quantity:

$$\text{STRESS} = \left[\frac{\sum_{i,j} (f(\delta_{ij}) - d_{ij})^2}{\sum_{i,j} d_{ij}^2} \right]^{1/2} . \qquad (6.1)$$

There are two types of MDS depending on the properties of $f(\delta_{ij})$: In the *metric MDS*, $f(\delta_{ij})$ is a monotonic, metric-preserving function. In the *non-metric MDS*, $f(\delta_{ij})$ is a weakly monotonic transformation; that is, it preserves only the rank ordering of the δ_{ij}'s. The monotonic transformation is computed via "monotonic regression" (also known as "isotonic regression") [14]. STRESS is a non-negative number that indicates how well dissimilarities are preserved in the embedding. For our purposes, when the STRESS is less than 0.2 the embedding is considered successful. Since only the relative dissimilarities are used, and not actual object coordinates, rigid transformations and reflections can be applied to the MDS result without changing the STRESS. Embedding methods such as SVD and PCA are not appropriate here because our signatures do not form a linear space, and we do not have the actual points, only the non-Euclidean distances between them. In our system we used the ALSCAL MDS program [98].

2. EXAMPLES

Figure 6.1 displays the result of a sample query returned by our color-based retrieval system as both a ranked list and as an MDS display. The query specified 20% blue and 80% "don't care" and requested the ten best matching images. In the MDS display, similar images of desert scenes with yellowish ground group together at the top left, images with green plants group at the bottom, and the two other images–a desert image with white ground and an image of a statue–are to the right. An all-blue image is comparatively dissimilar from the others, and is accordingly relegated to the far right. In the list, however, these relationships are not apparent.

In Figure 6.2 we see the MDS display of a query result where the user was looking for images of flowers. The top of the figure shows the specified amounts of red, pink, and green. The bottom of the figure is the MDS display (with STRESS=0.19) of the 20 best matches. One axis arranges the images from pink flowers to red flowers, while the other describes the shade of green in the images. Notice that image thumbnails placed at the coordinates returned by the MDS algorithm might occlude other thumbnails. Up to a point, this is not really a problem since these images are likely to be similar and are, therefore, well represented by the topmost thumbnail.

3. MISSING VALUES

As the number of images on which the MDS is applied increases, the computational time increases. When the EMD is used to compute the dissimilarity matrix, most of the computation time is taken by the computation of the EMDs. Table 6.1 shows the average time in seconds needed to compute the dissimilarity matrix and the MDS embedding for different number of images. These

Ranked list display

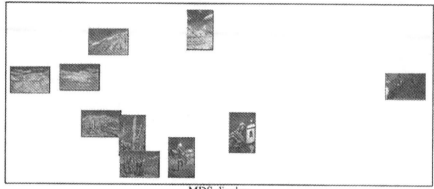

MDS display

Figure 6.1.   Looking for desert images; (a) Traditional display. (b) MDS display. *(This is a color figure)*

computations were done on a SGI Indigo 2 with a 250MHz CPU, using our color-based retrieval system.

number of images	10	20	50	100	200	500
dissimilarity matrix	0.02	0.12	0.64	2.62	11.02	71.33
MDS	0.01	0.02	0.08	0.33	1.78	25.92
total	0.03	0.14	0.72	2.95	12.80	97.25

Table 6.1.   Computation time in seconds for the dissimilarity matrix and the MDS.

One desirable property of the non-metric MDS is that if some dissimilarities are absent, they are simply left out of the formula for the STRESS (Equation (6.1)). This fact can be used to significantly reduce the computation time by not computing the full dissimilarity matrix. This is justified by the redundancy of information in the dissimilarity matrix which is due to the following two factors: the EMDs are metric so that the triangle inequality provides a large number of additional constraints. The second factor is the fact that for the non-metric MDS only the ranking of the dissimilarities matters and not their actual values. We found that using only twenty dissimilarities per image, chosen at random, results in a very similar MDS display to the one computed with the full dissimilarity matrix.

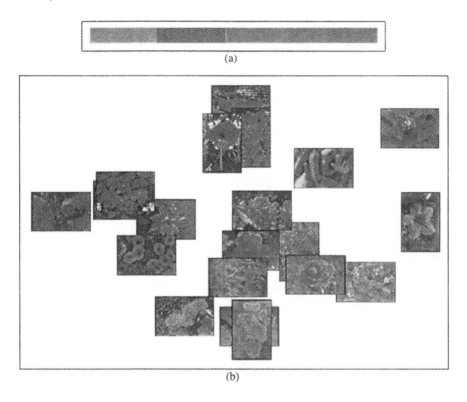

(a)

(b)

Figure 6.2. A color signature and the MDS display of the 20 best matches. *(This is a color figure)*

4. EVALUATING DISSIMILARITY MEASURES

In addition to creating a display to visualize query results, the MDS can be used as a tool for evaluating how well a dissimilarity measure agrees with perception. By applying MDS to a set of images, and displaying the results as we have done in the previous sections, it is easy to see if images that are perceptually similar are placed close to each other, and if dissimilar ones are far apart. Additional support can be obtained by comparing possible interpretations of the MDS axes to known results from psychophysics. Many psychophysical experiments study perceptual similarities, and present their results using MDS. In such cases a qualitative evaluation can be done by comparing the MDS display of a dissimilarity measure with the one obtained by a psychophysical experiment using the same images. In this section we show an example using MDS to evaluate the dissimilarity measures developed in Section 2 for homogeneous textures.

Figure 6.3 shows the results of applying a two-dimensional MDS to 16 homogeneous texture patches using the EMD with no invariance. For convenience, we repeat here the actual dissimilarity matrix that was computed in Table 4.1. In this figure[1], the hand-drawn arrows suggest that one axis reflects the coarseness of the texture, from fine to coarse. The other (curved) axis reflects the dominant orientation of the texture. On the left we see horizontal textures, on the right vertical textures, and as we move from left to right on the lower half-circle, the orientation changes counter clockwise. The other textures have no dominant orientation. STRESS in this figure is 0.063, so on average, dissimilarities in the picture are close to the real EMDs. We found that adding more oriented textures with different scales and orientations completes also the top half of the orientation circle, at the expense of the coarseness axis. This increases the STRESS, making two dimensions insufficient. In this case, a three-dimensional MDS would use two dimensions for the orientation circle and the third for coarseness.

A two-dimensional MDS display using the rotation-invariant EMD from Section 2.2 on the texture signatures in shown in Figure 6.4 (with a STRESS value of 0.077), together with its corresponding dissimilarity matrix. One axis emphasizes the directionality of the texture, where textures with one dominant orientation (any orientation) are at the top, and textures without a dominant orientation (no orientation at all, or more than one orientation) are at the bottom. The other axis is coarseness, similar to the previous experiment. An example of the rotational invariance can be seen by the two oriented fabric textures on the right that are close together although they have different orientations. Coarseness and directionality were found by psychophysical experiments by Tamura *et al.* [99] to be the two most discriminating texture properties for human perception.

Finally, we use the rotation- and scale-invariant EMD from Section 2.3. The two-dimensional MDS display, shown in Figure 6.5 (with STRESS equal to 0.1), can be interpreted as follows: One axis is again the directionality, while the other shows what we call the "scality" of the texture, a measure that distinguishes between textures with one dominant scale and textures with more than one, or no dominant scale. For example, the two textures of oriented bars, which have different orientations and scales, are close to each other when using the invariant EMD. Also, the two textures of tiles on the right are very close to each other even though they differ by more than three octaves in scale!

---

[1]In order to see the fine details of the textures better, only one-quarter of the textures in Figures 6.3-6.5 are displayed.

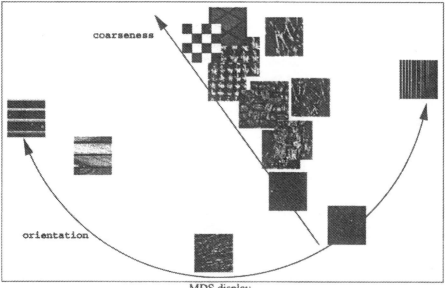

MDS display

	1	2	3	4	5	6	7	8	9	10	11	12	13	14	15	16
1	0.00															
2	0.39	0.00														
3	2.95	2.97	0.00													
4	1.82	1.96	2.35	0.00												
5	1.84	1.75	4.06	1.67	0.00											
6	1.07	1.14	2.08	2.77	2.89	0.00										
7	1.94	1.78	2.55	3.35	3.22	0.99	0.00									
8	1.10	0.92	2.26	2.84	2.65	0.39	1.07	0.00								
9	1.14	0.80	2.23	2.68	2.47	0.52	1.11	0.23	0.00							
10	1.00	0.74	3.07	2.70	2.24	1.11	1.74	0.90	0.89	0.00						
11	2.02	1.71	2.79	3.69	3.17	1.15	1.18	0.91	0.97	1.01	0.00					
12	0.94	1.07	3.38	1.57	1.68	1.91	2.88	1.86	1.76	1.55	2.56	0.00				
13	2.31	2.07	2.38	3.17	3.60	1.39	1.34	1.32	1.27	2.03	1.17	2.29	0.00			
14	3.84	3.86	0.80	3.23	4.92	2.86	2.62	3.05	3.11	3.95	3.17	4.22	2.13	0.00		
15	2.19	2.04	4.66	3.34	2.27	2.57	2.87	2.37	2.31	1.48	2.31	2.55	3.45	5.50	0.00	
16	1.51	1.48	1.96	2.96	3.17	0.76	1.19	0.76	0.82	1.55	0.93	2.32	0.88	2.41	3.08	0.00

Dissimilarity matrix

Figure 6.3. Two-dimensional MDS applied to 16 homogeneous textures using no invariance. The two dominant axes of orientation and coarseness emerge.

5. VISUALIZATION IN THREE DIMENSIONS

MDS can be computed in any number of dimensions, although only two and three dimensions are useful for visualization purposes. Going from two dimensions to three dimensions gives the MDS more flexibility in embedding the images, resulting in lower STRESS and a perceptually more pleasing display. We developed a system, based on the SGI Open Inventor object-oriented 3-D toolkit, that creates three-dimensional MDS displays, and allows the user

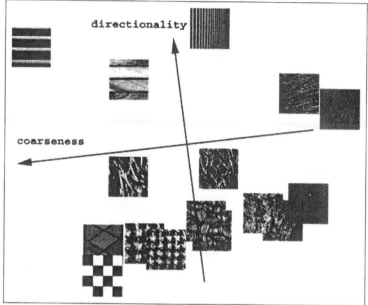

MDS display

	1	2	3	4	5	6	7	8	9	10	11	12	13	14	15	16
1	0.00															
2	0.23	0.00														
3	2.61	2.38	0.00													
4	1.40	1.46	1.80	0.00												
5	1.60	1.66	2.27	0.54	0.00											
6	1.07	0.93	1.69	2.25	2.45	0.00										
7	1.94	1.78	1.89	2.77	3.00	0.99	0.00									
8	1.08	0.92	1.67	2.27	2.47	0.28	1.07	0.00								
9	0.99	0.80	1.64	2.21	2.41	0.27	1.07	0.18	0.00							
10	0.93	0.74	1.70	1.38	1.62	0.96	1.64	0.90	0.89	0.00						
11	1.93	1.71	1.17	2.21	2.53	0.95	1.13	0.91	0.97	1.01	0.00					
12	0.80	0.94	3.19	1.53	1.54	1.67	2.13	1.82	1.73	1.51	2.56	0.00				
13	2.04	1.88	2.11	3.11	3.33	1.14	0.84	1.10	1.16	1.80	0.96	2.29	0.00			
14	3.50	3.27	0.80	2.68	3.14	2.48	1.97	2.47	2.53	2.58	1.58	4.04	1.86	0.00		
15	2.10	2.04	1.06	1.19	1.60	2.35	2.87	2.37	2.31	1.48	2.31	2.48	3.24	1.56	0.00	
16	1.51	1.44	1.93	2.96	3.17	0.70	1.06	0.69	0.72	1.53	0.90	2.19	0.82	2.41	3.07	0.00

Dissimillarity matrix

Figure 6.4. Two-dimensional MDS applied to 16 homogeneous textures with rotation invariance. The axes are coarseness and directionality.

to rotate, pan, and zoom in space. Since the viewing direction can be arbitrary, we texture-map the thumbnails onto the six faces of a cube and position these cubes in the appropriate locations in space. A significantly larger number of images can be conveniently embedded in three-dimensional space than in the two-dimensional case. Figure 6.6 shows two different viewpoints of an MDS display applied to 500 random images. The dissimilarity matrix was computed

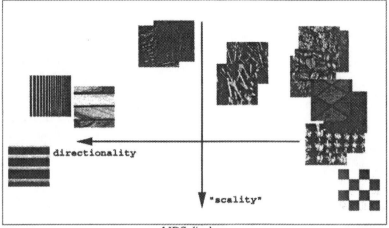

MDS display

	1	2	3	4	5	6	7	8	9	10	11	12	13	14	15	16
1	0.00															
2	0.23	0.00														
3	1.98	1.91	0.00													
4	1.40	1.46	0.94	0.00												
5	1.39	1.33	0.92	0.54	0.00											
6	0.55	0.54	1.69	1.62	1.50	0.00										
7	1.27	1.22	1.89	1.76	1.83	0.97	0.00									
8	0.53	0.50	1.67	1.62	1.46	0.28	1.07	0.00								
9	0.50	0.48	1.64	1.54	1.43	0.27	1.01	0.18	0.00							
10	0.93	0.74	1.33	0.89	0.90	0.96	1.27	0.90	0.89	0.00						
11	1.18	1.01	1.17	0.97	1.02	0.95	1.13	0.91	0.97	0.35	0.00					
12	0.80	0.94	2.08	1.53	1.54	0.92	0.83	1.04	0.94	1.02	0.96	0.00				
13	0.90	0.96	1.95	1.69	1.63	0.91	0.84	0.83	0.77	1.12	0.96	0.41	0.00			
14	2.44	2.38	0.49	1.34	1.42	2.14	1.97	2.11	2.05	1.91	1.58	2.24	1.86	0.00		
15	2.10	2.04	0.26	1.10	1.06	1.97	2.19	1.96	1.96	1.48	1.39	2.22	2.22	0.45	0.00	
16	0.87	0.84	1.93	1.86	1.79	0.70	1.06	0.69	0.72	0.99	0.90	0.74	0.82	2.41	2.21	0.00

Dissimilarity matrix

Figure 6.5. Two-dimensional MDS applied to 16 homogeneous textures with rotation and scale invariance. The two axis are "scality" and directionality.

using the EMD and the color signatures. Images were organized in space by their dominant colors, and by their brightness. Notice that similar images are close to each other, and the user can find images of sunsets, images of green fields with blue skies, or images taken at night at a glance. The interpretation of the axes varies according to the set of images; for example, Figure 6.7(a) is the result of the same query as used in Figure 6.1 but with 500 images and a three-dimensional MDS display. One axis can clearly be interpreted as the amount of blue in the image, while the other two are the brightness and the amount of yellow. Part (b) of the figure zooms into the part of space that contains images of deserts.

Figure 6.6.   Two views of a three-dimensional MDS display applied to 500 random images. Only color information was used. *(This is a color figure)*

(a)

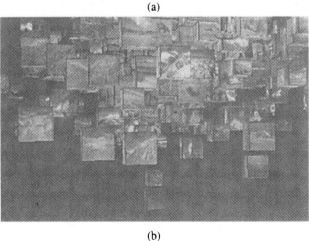

(b)

Figure 6.7. Looking for images with 20% blue. (a) The 500 most relevant images. (b) Focusing on images of deserts. *(This is a color figure)*

Although a three-dimensional MDS display conveys more information than its two-dimensional counterpart, we found that the latter is more useful for image retrieval, as rendering a three-dimensional MDS display takes significantly more time. In addition, since only two-dimensional projections can be displayed on the screen, the user needs to rotate the three-dimensional MDS display in order to understand its structure. By the time the three-dimensional structure is rendered and understood, the user can go through a series of two-dimensional MDS displays, which can be comprehended at a glance, thereby surpassing the advantages of the three-dimensional MDS.

6.   SUMMARY

We showed that by using the mutual dissimilarities between a set of images, the images from a query result can be displayed in a more intuitive way than with the traditional methods. The resulting embeddings, in two- and three-dimensions, are adaptive and reflect the axes of maximum variation. A larger number of images can displayed, helping the user to form a global mental model of the entire database. In addition, it is easier to understand what are the commonalities is the returned set of images and where to go next, as well as when to stop the search.

The idea of an adaptive image embedding can be applied to other modalities besides color and texture, as long as some notion of similarity, metric or not, continuous or discrete, can be defined.

In the next chapter we extend this visualization technique to allow for intuitive navigation through large collections of images for the purpose of image retrieval.

Chapter 7

NAVIGATION

We're not lost. We're locationally challenged.

—John M. Ford

The MDS-based visualization technique, described in the last chapter, makes it possible to display a large number of images in an intuitive way that allows the user to see at a glance the commonalities is the returned set of images and what part of the display is most relevant. In this chapter, we develop a navigation method based on this display. Once the user points to the region of interest on the display, the system zooms in and finds more images that are similar to the common features of the images in the region. By iterating this refinement process, the user can quickly home in to the relevant parts of the database. We show examples of navigation in a space of color images and in a space of police mugshots.

1. RETRIEVAL BY NAVIGATION

The discrepancy that results from using syntactic features, such as color and texture, to satisfy semantic queries causes a basic problem with the traditional query/response style of interaction. An overly generic query yields a large jumble of images, which are hard to examine, while an excessively specific query may cause many good images to be overlooked by the system. This is the traditional trade-off between good precision (few false positives) and good recall (few false negatives). Striving for both good precision and good recall may pose an excessive burden on the definition of a "correct" measure of image similarity. While most image retrieval systems recognize this and allow for an iterative refinement of queries, the number of images returned for a query is usually kept low so that the user can examine them, one at a time, in order of similarity to the query.

If, as we showed in Chapter 6, images are arranged on the screen so as to reflect similarities and differences between their feature distributions, the initial queries can be very generic, and return a large number of images. The consequent low initial precision is an advantage rather than a weakness, as it provides a neighborhood of images the user can relate to when refining the query and deciding where to go next. In fact, the user can see large portions of the database at a glance, and form a mental model of what is in it. Rather than following a thin path of images from query to query, as in the traditional approach, the user now *zooms in* to the images of interest. Precision is added incrementally in subsequent query refinements, and fewer and fewer images need to be displayed as the desired images are approached. This can be envisioned as analogous to visiting a bookstore. Books are shelved according to their subjects, for example, travel-books, cookbooks, poetry, etc. After locating the right section, say cookbooks, the books are again classified into different categories such as vegetarian cooking, baking, etc. The books can be further sorted by title or by author. The last step when searching for a book is to scan the relevant set of books, one at a time, and find the desired one. After a few visits to the bookstore, a global mental picture of it is formed, with the knowledge of what kinds of books can be found and where to look for them.

When the user selects the region of interest on the display, a new, more specific query is automatically generated, reflecting the common features of the images in the selected set. A new set of images is returned and displayed by a new MDS, which now reflects the new dominant axes of variation. By iterating this process, the user is able to quickly navigate to the portion of the image space of interest, typically in very few mouse clicks.

This navigation scheme can be categorized as a *relevance feedback* system [81, 59, 92] where retrieval is guided by feedback from the user, who marks relevant and non-relevant images. A major difference from other relevance feedback systems is that in our navigation approach, the relevant images are likely to be displayed close to each other, and far from ones which are completely irrelevant. This makes the selection of the relevant set easier and more intuitive, as the user naturally moves his focus to relevant parts of the display, while in the other methods the user needs to scan the list of images, one at a time, and decide whether it is relevant or not.

We propose a navigation scheme in which the user starts by specifying a generic query. Using partial queries is encouraged when the user is not confident about certain features by specifying "don't care" rather than guessing. A large set of images is returned and embedded in two-dimensional space using MDS. Once the user selects an area of interest, an appropriate new query is automatically generated and submitted. Now a smaller number of images is returned by the system. The new set of images is not necessarily a subset of the previous set, so images which were not returned by the previous query can still be retrieved at later stages, as the query becomes more precise. A new MDS display is shown, with new axes of variation based on the new image set.

To generate the new query, our user interface allows the user to select from the MDS display images that will be used to form the next query. The selection can consist of one or multiple regions. The signatures of the selected k images are combined to generate a signature that will be used as the next query. We want the new signature to reflect features which are common in these images, and ignore the others by using "don't care". This is done by representing the clusters of all the signatures in the return set as points in the feature space and finding the dominant clusters in that space. We use a similar clustering algorithm as the one that was used in Section 2 to generate the color signatures. We reject clusters that are not represented by points from at least $\lceil k/2 \rceil$ images to insure commonality. Each new cluster is assigned a weight that is the median of the weights of the points in that cluster. We use the median instead of the mean for robustness with respect to points that have very different weights. Finally, we normalize the resulting signature by dividing the weights of the clusters by the number of points that contributed to it. If the total weight of the signature is now greater than 100%, we normalize it again. Usually, however, the total weight of the signature will be less than 100%, so the reminder is marked "don't care."

2. AN EXAMPLE

An example of retrieval by navigation using color signatures is given in Figures 7.1 and 7.2. Suppose we are looking for images of skiers. These images can be characterized by blue skies and white snow, so we use as our query "20% blue, 20% white and 60% 'don't care'." The query is illustrated in Figure 7.1(a). Since this is a generic query, the precision is bound to be low as confirmed by the list of ten images in Figure 7.1(b). Only one of the ten images is relevant (although the color signatures of the others matches the query well), and, in general, consecutive images in the list can be very different from each other. Figure 7.1(c) displays the MDS embedding of the best one hundred matches. Although a large number of images is displayed, it is easy to see their structure. Images of underwater scenes are at the left, images with plants are at the top right, and so forth. The desired images of skiers are clearly at the right. At this point the user selects, with the mouse, the relevant region as shown by the black ellipse in Figure 7.1(c). The new generated query, illustrated in Figure 7.2(a), resulted in the images shown in Figures 7.2(b)-(c). Note that blue and white are the dominant colors in the images of the skiers, although not the same shades as specified in the original query. The precision was significantly improved, and nine out of the ten best matches are now images of skiers. Figure 7.2(c) shows

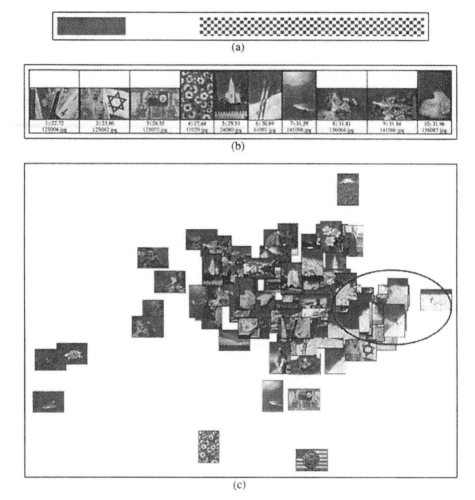

Figure 7.1. Retrieving skiers: first iteration. (a) The query's signature: 20% blue, 20% white, and 60% "don't care" (represented by the checkerboard pattern). (b) The 10 best matches sorted by similarity to the query. (c) MDS display of the 100 best matches (STRESS=0.15), where the relevant region is selected. *(This is a color figure)*

the MDS embedding of the best twenty matches; the axes of maximal variation are the amount of blue and the brightness. Images with more blue are at the top of the display, while brighter ones are at the right.

3.   PROCRUSTES ANALYSIS

A typical retrieval-by-navigation session involves a series of MDS embeddings, each a refined version of its predecessor. Consecutive embeddings along this path are likely to contain images that appear in more than one embedding.

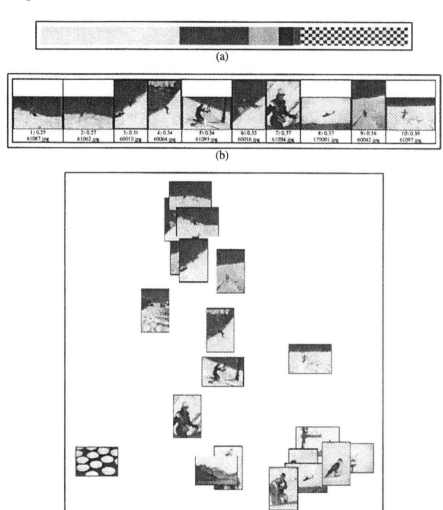

Figure 7.2. Retrieving skiers: second iteration. (a) The generated signature. (b) The 10 best matches sorted by similarity to the query. (c) MDS of the 20 best matches (STRESS=0.06). *(This is a color figure)*

In order not to disorient the user, we would like these images to be positioned in the new embedding as close as possible to their positions in the old one. However, the MDS can be computed only up to a rigid transformation, which can cause successive MDS embeddings to be translated, rotated, scaled, and flipped relative to previous ones. Procrustes analysis [14, 34] can be used to align one MDS embedding with respect to another. In general, Procrustes analysis computes the optimal similarity transformation that must be applied to one embedding in order to match it, as well as possible, to the other. Such alignment is applied only to images that are in the intersection of the two image sets.

Assume that we are given two consecutive MDS embeddings where n images participate in both. Let $\mathbf{X} = [\mathbf{x}_1, \dots, \mathbf{x}_n]^T$ be an n-by-2 matrix representing the two-dimensional coordinates of the n images in the first embedding. Similarly $\mathbf{Y} = [\mathbf{y}_1, \dots, \mathbf{y}_n]^T$ represents the coordinates of the n images in the second embedding. In general, the two embeddings can be in spaces with different dimensions.

We measure the discrepancy between the two embeddings by

$$D = \sum_{r=1}^{n} (\mathbf{y}_r - \mathbf{x}_r)^T (\mathbf{y}_r - \mathbf{x}_r) ,$$

Our goal then is to find the translation $\mathbf{b}$, the dilation ρ, and the orthogonal matrix $\mathbf{R}$ that will transform the points $\mathbf{y}_r$ to

$$\mathbf{y}'_r = \rho \mathbf{R}^T \mathbf{y}_r + \mathbf{b} ,$$

such that D will be minimized. The optimal translation and dilation are determined by our desire to best fill the screen display area. The optimal orthogonal matrix $\mathbf{R}$ can be found as follows [90]: we want to minimize

$$
\begin{aligned}
D &= \sum_{r=1}^{n} (\mathbf{R}^T \mathbf{y}_r - \mathbf{x}_r)^T (\mathbf{R}^T \mathbf{y}_r - \mathbf{x}_r) \\
&= \sum_{r=1}^{n} \mathbf{y}_r^T \mathbf{y}_r + \sum_{r=1}^{n} \mathbf{x}_r^T \mathbf{x}_r - 2 \sum_{r=1}^{n} \mathbf{x}_r^T \mathbf{R} \mathbf{y}_r \\
&= \text{tr}(\mathbf{Y}\mathbf{Y}^T) + \text{tr}(\mathbf{X}\mathbf{X}^T) - 2\text{tr}(\mathbf{X}\mathbf{R}\mathbf{Y}^T) ,
\end{aligned}
$$

where $\text{tr}(\cdot)$ is the trace of a square matrix.

Since only $\mathbf{R}$ is a variable, we need to maximize $\text{tr}(\mathbf{X}\mathbf{R}\mathbf{Y}^T)$. We can write

$$\text{tr}(\mathbf{X}\mathbf{R}\mathbf{Y}^T) = \text{tr}(\mathbf{R}\mathbf{Y}^T\mathbf{X}) = \text{tr}(\mathbf{R}\mathbf{C}) ,$$

where $\mathbf{C} = \mathbf{Y}^T\mathbf{X}$. Let $\mathbf{C}$ have the singular value decomposition, $\mathbf{C} = \mathbf{U}\mathbf{\Lambda}\mathbf{V}^T$, then

$$\text{tr}(\mathbf{R}\mathbf{C}) = \text{tr}(\mathbf{R}\mathbf{U}\mathbf{\Lambda}\mathbf{V}^T) = \text{tr}(\mathbf{V}^T\mathbf{R}\mathbf{U}\mathbf{\Lambda}) .$$

Since $\mathbf{R}$, $\mathbf{U}$, and $\mathbf{V}$ are orthonormal, and an orthogonal matrix cannot have any element greater than one, we get

$$\text{tr}(\mathbf{R}\mathbf{C}) = \text{tr}(\mathbf{V}^T\mathbf{R}\mathbf{U}\mathbf{\Lambda}) \leq \text{tr}(\mathbf{\Lambda}) .$$

Thus, D is minimized if $\text{tr}(\mathbf{R}\mathbf{C}) = \text{tr}(\mathbf{\Lambda})$, which is true when

$$\mathbf{V}^T\mathbf{R}\mathbf{U}\mathbf{\Lambda} = \mathbf{\Lambda} .$$

This can be written as

$$\mathbf{RU\Lambda V}^T = \mathbf{V\Lambda V}^T ,$$

so

$$\mathbf{RC} = \mathbf{V\Lambda V}^T = (\mathbf{V\Lambda^2 V}^T)^{\frac{1}{2}} = (\mathbf{V\Lambda UU}^T\mathbf{\Lambda V}^T)^{\frac{1}{2}} = (\mathbf{C}^T\mathbf{C})^{\frac{1}{2}} .$$

When $\mathbf{C}$ is nonsingular, the optimal orthogonal matrix $\mathbf{R}$ is then

$$\mathbf{R} = (\mathbf{C}^T\mathbf{C})^{\frac{1}{2}}\mathbf{C}^{-1} .$$

4. NAVIGATING IN A SPACE OF POLICE MUGSHOTS

This section describes a police mugshot retrieval system that was developed as a feasibility test for a Canadian police department, using the navigation ideas described in this chapter.

In order to identify a suspect in a crime, witnesses are often asked to scan large collections of police mugshots. Fatigue and insufficient attention span can lead to distraction during this process. A system that lets witnesses *navigate* through the mugshot collection in a more coherent fashion can help reduce the likelihood of costly mistakes.

The witness would give general indications about the appearance of the suspect, such as age group, sex, race, and hair color. Some of these attributes are inequivocal, so an initial selection process can rule out irrelevant pictures. Other attributes, such as hair or skin color, are harder to categorize, and a yes/no answer would not be adequate. Instead, the system can display many relevant images as small thumbnail icons on a single screen, arranging them in such a way that similar faces, in terms of the attributes of importance for the given search, appear close to each other on the screen.

MDS allows for the automatic generation of such displays. Because similar images are nearby, it becomes much easier for a witness to concentrate his or her attention on the part of display of interest. By selecting an "interesting" part of the display, the system produces a new display, with images that are similar to those in the interesting part. By repeating this procedure, the witness can home in to the image of the suspect in a few steps.

To illustrate our approach, Figure 7.3 shows a display of a set of 200 mugshots. This display was computed automatically from the annotated images (more on annotations below). The system discovers autonomously that faces can be cleanly separated into male and female, where men with mustaches and beards are further away from women than men with no facial hair. Skin color appears as a second dominant feature, and the distribution of images on the display reflects different gradations of skin color. One diagonal axis separates men from women, and the orthogonal axis separates white people from black people with "brown" people in between. We emphasize that these axes were not predefined, but were, in a sense, discovered by the system (although gender and race are among the hand–annotated features that are used to compute the dissimilarity).

Figure 7.3. Navigation 1: gender vs. skin color. The mugshots are intentionally blurred for anonymity. *(This is a color figure)*

Clicking in the group of black men gave the 40 mugshots shown in Figure 7.4. Now the interpretation is different: young vs. old, and thin vs. fat. Again, the interpretations of the axes were found automatically. As we zoom in, the interpretations will consist of finer details, maximizing the variability of the mugshot subset at hand.

In order to apply our perceptual navigation algorithms, we needed a set of images and a similarity measure between them. The image set was provided by a Canadian police department. We used a very simple feature-based similarity measure between mugshots, based on the simple, pre-annotated features shown in Table 7.1 together with their possible values.

We defined a distance measure for every feature, and the similarity measure between two mugshots was defined as a linear combination of these distances. The relative importance of the features was controlled by modifying the weights of each feature. The user could also "turn off" features which should not participate in the computation of the similarity measure.

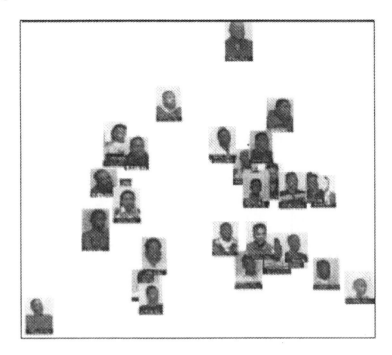

Figure 7.4. Navigation 2: age vs. weight. The mugshots are intentionally blurred for anonymity. *(This is a color figure)*

feature	values
skin color	white, brown, black
gender	male, female
color	black, brown, blond, red, white
hair style	long, medium, short, bald
beard	yes, no
mustache	yes, no
eye color	brown, blue, green
glasses	yes, no
weight	1 (thin), ..., 6 (fat)
age	1 (young), ..., 6 (old)

Table 7.1. The appearance-based mugshot features used for retrieval.

5. A DEMONSTRATION SYSTEM

Combining the principles from the previous chapters leads to the navigation-based retrieval scheme sketched in Figure 7.5. The user starts with a generic query with the goal to get to the approximate neighborhood of the database.

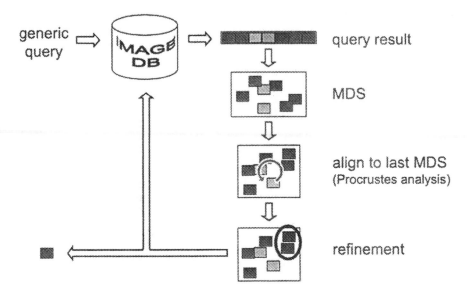

Figure 7.5.   Retrieval by navigation. *(This is a color figure)*

The retrieval system returns a list of images on which an MDS is computed. In subsequent queries, the MDS results are rotated, using Procrustes analysis to match, as well as possible, the previous embeddings. The user can refine the query by selecting the set of the most relevant images from the MDS display, which is used to compute the next query. This navigation cycle is iterated until the desired images are found, or until the user decides to terminate the search.

We developed a demonstration system for color-based image retrieval[1]. The system uses our 20,000-image database and runs on the World-Wide-Web (WWW). A snapshot of the system is shown in Figure 7.6.

The screen is divided into three main frames:

- *Query input* (the top-left part of the screen): Here the user can sketch a query by coloring cells in the 5-by-5 query grid. There are 25 cells, so each one corresponds to 4% of an image. The colors are picked using a color wheel (implemented as a Java applet), based on the HSV color space (see Section 1.1). The query grid defines the color signature that is used as the query. In addition, the user can specify the number of images the system should return, and other parameters such as whether to compute the full distance matrix for the MDS or use the missing values option.

---

[1]The demo can be accessed through `http://vision.stanford.edu/~rubner`.

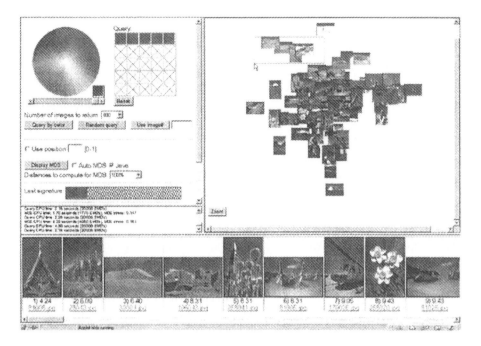

Figure 7.6. A snapshot from the color-based retrieval demonstration system. *(This is a color figure)*

- *Query result* (the bottom of the screen): This frame displays a scrollable list of images returned by the last query, sorted by their distances from the query. The distances and the image numbers are shown together with a small version of the image. Clicking on the image initiates a new search where the color signature of that image is used as the query. Clicking on the image number opens a window with the full size image.

- *MDS map* (the top-right of the screen): Here the results of the MDS embeddings are shown using small image thumbnails. Using the mouse, the user can explore occluded images. The mouse can also be used to select sets of images which generate the next query. Such a query is generated when the user clicks the "zoom" button.

The system supports the following queries:

1 *Random query*: a specified number of random images is returned.

2 *Query-by-sketch*: the user explicitly specifies the color signature.

3 *Query-by-example*: the color signature of an image is used for the query.

4 *Query-by-refinement*: the query is generated based on the common features in a selected set of images.

The example in Figure 7.6 shows the first step of a search where the user is looking for images of deserts. The initial generic query is 20% blue for the sky and 80% "don't care." The best matches are shown in the results frame, and the MDS embedding of 100 returned images is shown in the MDS frame. The desert images, in the top left, are selected for the next query.

Currently, our demonstration system satisfies a query by using a linear search over all the images in the database, comparing their color signatures to the query, and keeping the N best matches, where N is specified by the user. When the total weight of the query is equal to the weights of the signatures, the lower bound from Section 3.2 is used and the EMD between the signature and the current image signature is computed only when the lower bound between the two is less than the largest distance between the query and the N best matches so far. To be used as a full image retrieval system, this demonstration system should be extended to deal with better indexing, such that the image descriptors are stored in a way that allows a sub-linear retrieval time. In addition, it should combine other modalities beside color, including texture, shape, and composition, and should support a query language were boolean operations are supported.

6.   SUMMARY

Intuitive image retrieval is a hard and ill-defined problem. Our attempt toward a solution combines the EMD as a dissimilarity measure that agrees with perceptual similarity, MDS as a better display technique that is easy to comprehend, and a refinement process that automatically generates successive queries. This framework reduces the need to directly deal with image features, as queries can be posed by the user starting with a vague description or even a random set of images and then zooming into the relevant parts of the display. Our approach works well for color-based and police mugshot retrieval.

Chapter 8

CONCLUSION AND FUTURE DIRECTIONS

There will come a time when you believe everything is finished. That will be the beginning.
—Louis L'Amour

1. CONCLUSION

The methods presented in this book provide a novel set of tools and opens real possibilities for content-based image retrieval systems. Tools are provided for image retrieval, visualization of query results, and navigation in the space of images.

In the introduction, we listed questions that were to be addressed in this book. We now reconsider these questions, each followed by our approach and contributions towards answering it:

- *What features describe the content of an image well?*

  In Chapters 3 and 4 we focused on color and texture features, respectively. For both, we carefully chose appropriate feature spaces and ground distances so that the distances between single features agrees with perception. We showed that these features lead to effective retrieval.

  For color, we used the CIELab space, which was designed so that distances between single colors conform to perceptual similarity. We presented an extension of the color features that also included the absolute position of colors within the image. We used this extension when the layout of colors in the image was important. For texture, we designed Gabor filters on a log-polar grid to model the texture content of an image. We developed dissimilarity measures for homogeneous textures and for images that comprise many textures. For homogeneous textures we also presented dissimilarity measures invariant to changes in rotation and scale. In Section 4 we showed that these texture features lead to dissimilarity measures that agree with perception.

- *How to summarize the distribution of these features over an image?*

  In Chapter 1 we introduced signatures as a compact and efficient representation for feature distributions. We compared signatures to the commonly used histograms and showed, in Chapter 5, that signatures often lead to better results than histograms, even when the size of the signatures is significantly smaller than that of the histograms.

- *How do we measure the dissimilarity between distributions of features?*

  The core of any retrieval system is a dissimilarity measure that is able to find and return the most relevant images in a way that conforms to the users' notion of similarity. We made a distinction between the ground distance, which measures dissimilarity of single features, and the dissimilarity between feature distributions. In Chapter 2 we developed the EMD to extend the ground distance to dissimilarities between feature distributions. The EMD is a general and flexible metric and does not suffer from the binning problems of most extant definitions of distribution dissimilarity. It allows for partial matches, and lower bounds are readily available for it. In addition, as demonstrated in Chapter 5, for images with moderate similarities, the EMD gave better results than other dissimilarity measures.

- *How can we effectively display the results of a search?*

  Instead of displaying the returned images as a list sorted by dissimilarity from the query, as done in most image retrieval systems, we use the full dissimilarity matrix of the images. We showed in Chapter 6 that applying MDS to the dissimilarity matrix and placing image thumbnails at the coordinates of the resulting low-dimensional embeddings leads to an intuitive display that conveys information about how the images relate to each other.

- *How can a user browse the images in the database in an intuitive and efficient way?*

  We coupled the notion of browsing with that of visualization. No browsing is possible without an appropriate display. Using our display technique we developed a navigation scheme that allows the user to intuitively retrieve images by zooming in promising areas of the display, with refinement of the query done automatically. We demonstrated this technique for color-based retrieval and for police mugshot retrieval.

Additional contributions of this book are the following:

- In Chapter 5 we provided an extensive comparison of various dissimilarity measures that are used in image retrieval systems. Comparisons were made for different combinations of parameters, such as the number of bins (for the histogram-based dissimilarity measures), the size of the filter bank (for texture), and the balance between precision and recall.

- We presented texture-based retrieval without the need to segment the images first as dome by most methods. This is possible due to the combination of the texture features with the ability of the EMD to do partial matches.

- We developed a demonstration system, described in Section 5, by which people can access remotely and experiment with the methods presented in this book on a 20,000-image database.

2. FUTURE DIRECTIONS

The EMD can be applied to other modalities besides color and texture as long as a ground distance can be defined in the appropriate feature space. Examples include shape, compositions of objects, eigenimage similarity, among other image features [3, 24, 29, 64, 66, 85, 96]. A larger ensemble of features from different modalities can improve the overall performance of an image retrieval system. The EMD can benefit other applications beside image retrieval. Good results were achieved for the comparison of vector fields [51] and for detecting color edges [83].

The idea of adaptive image embedding is also not limited to color and texture and can be applied to other modalities as long as some notion of similarity, metric or not, continuous or discrete, can be defined. This method was successfully applied to computer-assisted parameter setting for animation [58], and to visualization of vector field images [51].

In this context, the key question, which we leave for future work, is to determine whether the embeddings and the main axes of variation "discovered" by MDS for each of these distances and for various types of image distributions are perceptually meaningful. We believe that since MDS groups similar images together—and away from dissimilar ones—this is often the case. We also plan to study more the relations between the axes chosen by MDS for related or overlapping image sets. Knowing the correspondence between these 'local charts' (in the sense of topology) of the image space may help provide a better sense of navigation.

Although our experience is that the process of image retrieval is improved by our navigation scheme, a more principled comparison between database navigation techniques is needed. Current research that studies the responses of users for different image retrieval methods [73] reports that most users find our navigation method effective[1].

---

[1]The experiment in [73] used our navigation method with only color information. The users, in general, agreed that although the method is promising, color alone is not sufficient for image retrieval.

A query language that allows more complex queries must be devised for effective image retrieval. Such a language needs to include operators like "and," "or," and "not." In addition, image retrieval systems need to integrate different modalities such as color, texture, and shape. The correct combination of these features, and the adjustment of their relative importance according to the user responses, is another topic of future research.

Appendix A
The Standard Deviation of the Gabor Filters

In this appendix, we derive the standard deviations of the Gabor filters in the radial (σ_u) and angular (σ_v) axes. These values are used in Equation (4.3).

1.   THE RADIAL STANDARD DEVIATION

For the radial axis, consider the half-amplitude contours of the filters with orientation $\theta = 0$ as in Figure A.1, where we require neighboring contours to touch each other.

Denote the half-width of the smallest filter along the u-axis by t, and the ratio of half-widths of two consecutive filters by a. Summing up the half-widths between U_l and U_h gives

$$
\begin{aligned}
U_h - U_l &= t + 2at + 2a^2 t + \ldots + 2a^{M-2}t + a^{M-1}t \\
&= 2t \sum_{m=0}^{M-1} a^m - t - a^{M-1}t \\
&= 2\frac{1 - a^M}{1 - a}t - (1 + a^{M-1})t \\
&= \frac{a+1}{a-1}(a^{M-1} - 1)t \ .
\end{aligned}
$$

The half-amplitude of a Gaussian with standard deviation σ is $\sigma\sqrt{2\ln 2}$. In our case, the half-amplitude of the largest filter can be written as

$$
a^{M-1}t = \sigma_u \sqrt{2\ln 2} \ .
$$

Using the above equations together with $U_h = a^{M-1}U_l$ leads to

$$
\sigma_u = \frac{a-1}{a+1}\frac{U_h}{\sqrt{2\ln 2}} \ .
$$

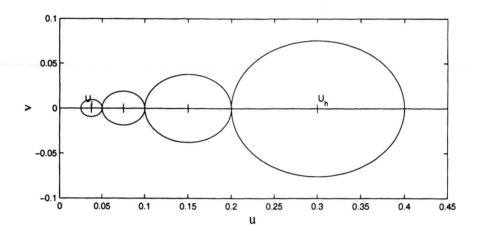

Figure A.1.   Half-amplitude contours of the filters with orientation $\theta = 0$.

2.   THE ANGULAR STANDARD DEVIATION

To derive σ_v, we plot the half-amplitude contours as in Figure A.2. Note that two neighboring ellipses meet on the line halving the angle $\phi = \frac{\pi}{L}$, where L is the number of orientations. Finding σ_v is thus accomplished by finding the value for which a line with slope $\frac{\phi}{2}$ is tangent to the ellipse

$$\frac{(u - U_h)^2}{2 \ln 2 \sigma_u^2} + \frac{v^2}{2 \ln 2 \sigma_v^2} = 1 \, ,$$

or

$$(u - U_h)^2 \sigma_v^2 + v^2 \sigma_u^2 = 2 \ln 2 \sigma_u^2 \sigma_v^2 \, .$$

Setting $v = \tan \frac{\phi}{2} u$ gives

$$(u - U_h)^2 \sigma_v^2 + \tan^2 \frac{\phi}{2} \sigma_u^2 = 2 \ln 2 \sigma_u^2 \sigma_v^2 \, ,$$

$$(\sigma_v^2 + \tan^2 \frac{\phi}{2} \sigma_u^2) u^2 - 2 U_h \sigma_v^2 u + U_h^2 \sigma_v^2 - 2 \ln 2 \sigma_u^2 \sigma_v^2 = 0 \, .$$

This equation has only one solution (tangent case), if

$$4 U_h^2 \sigma_v^4 - 4 (\sigma_v^2 + \tan^2 \frac{\phi}{2} \sigma_u^2)(U_h^2 \sigma_v^2 - 2 \ln 2 \sigma_u^2 \sigma_v^2) = 0 \, ,$$

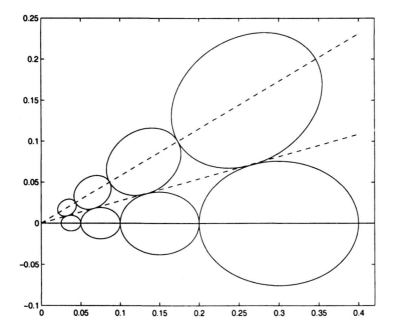

Figure A.2. Half-amplitude contours of two neighboring filters for different orientations.

which leads to

$$\sigma_v = \tan\frac{\phi}{2}\sqrt{\frac{U_h^2}{2\ln 2} - \sigma_u^2}\ .$$

Appendix B
Retrieval Performance Plots

In this appendix we collect plots of the results of the experiments we conducted in Chapter 5. In these plots, the EMD for signatures is always computed using 8 clusters.

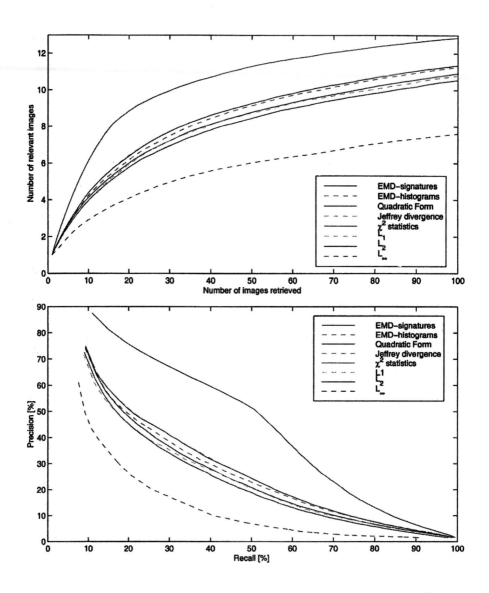

Figure B.1.   Retrieving color distributions using 32 bin histograms. *(This is a color figure)*

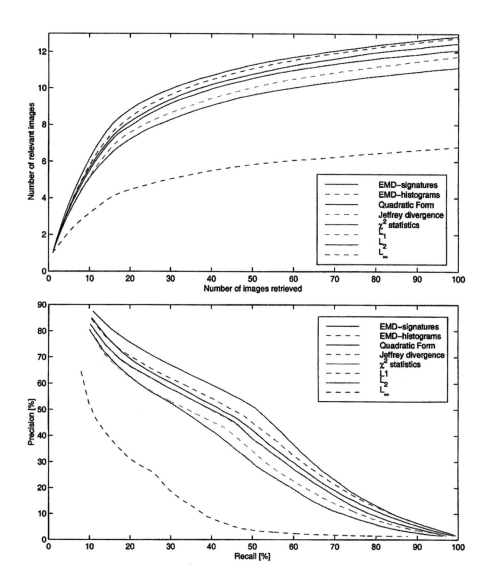

Figure B.2. Retrieving color distributions using 128 bin histograms. *(This is a color figure)*

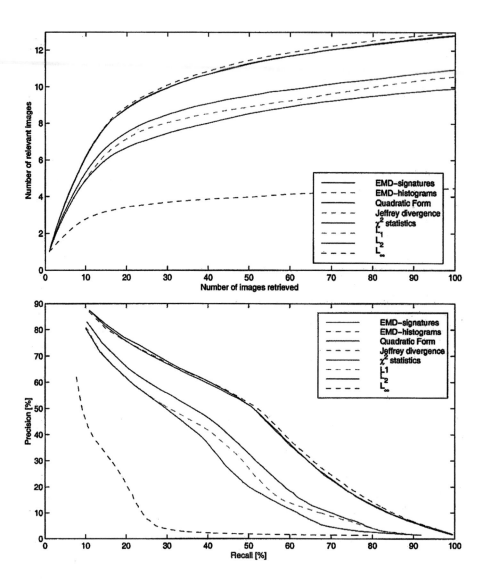

Figure B.3.   Retrieving color distributions using 512 bin histograms. *(This is a color figure)*

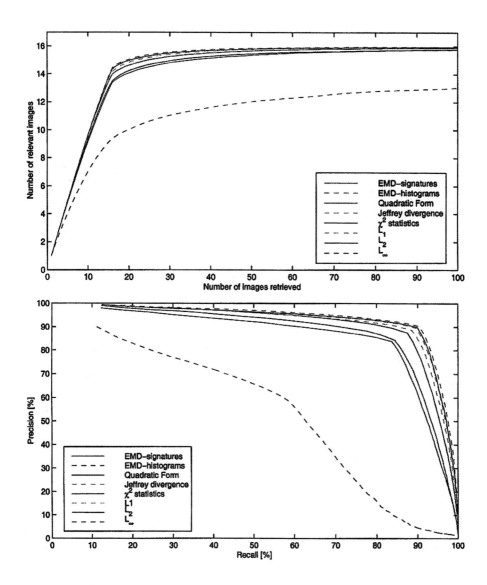

Figure B.4. Retrieving color distributions using 512 bin histograms for non-moderate similarities. *(This is a color figure)*

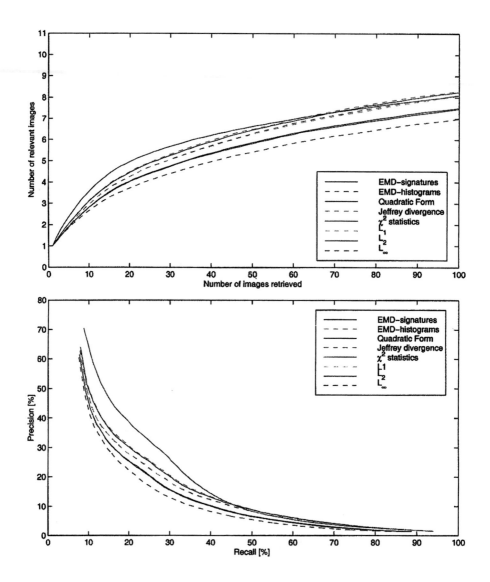

Figure B.5.   Retrieving texture distributions using 12 filters and 32 bin histograms. *(This is a color figure)*

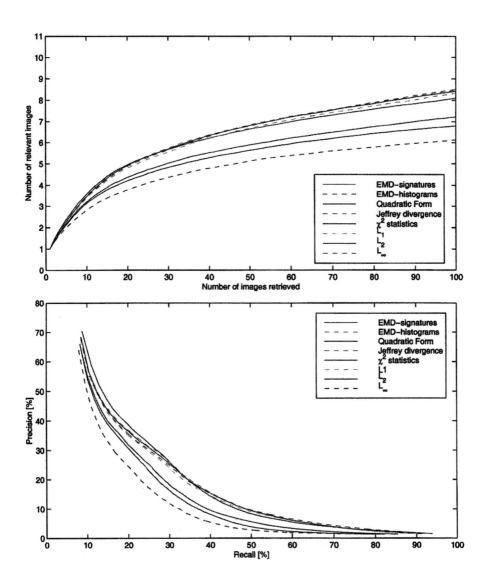

Figure B.6. Retrieving texture distributions using 12 filters and 128 bin histograms. *(This is a color figure)*

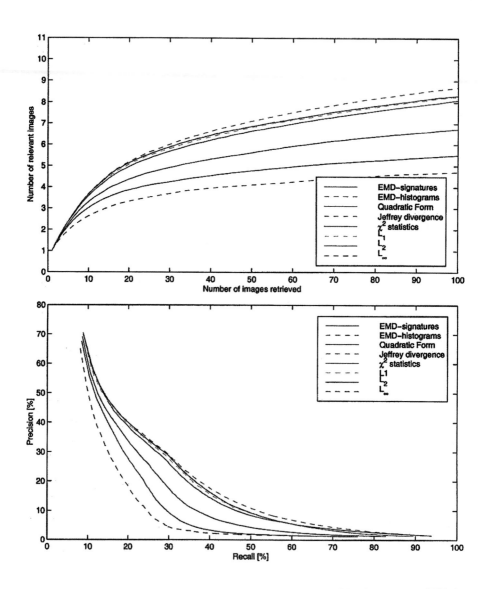

Figure B.7.   Retrieving texture distributions using 12 filters and 512 bin histograms.. *(This is a color figure)*

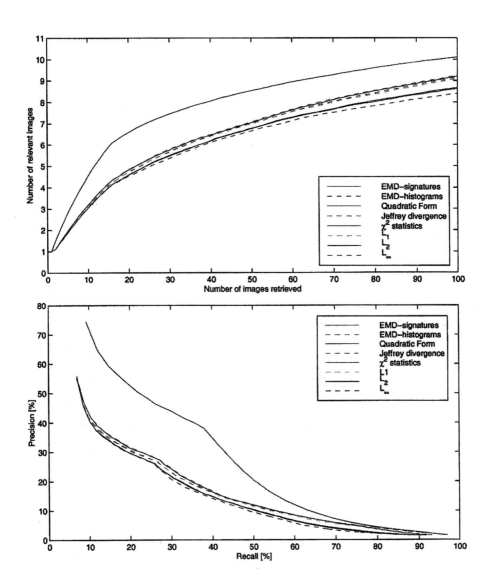

Figure B.8. Retrieving texture distributions using 24 filters and 32 bin histograms. *(This is a color figure)*

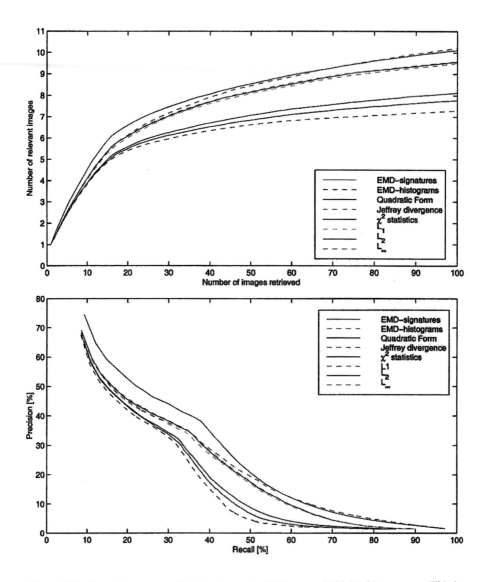

Figure B.9. Retrieving texture distributions using 24 filters and 128 bin histograms. *(This is a color figure)*

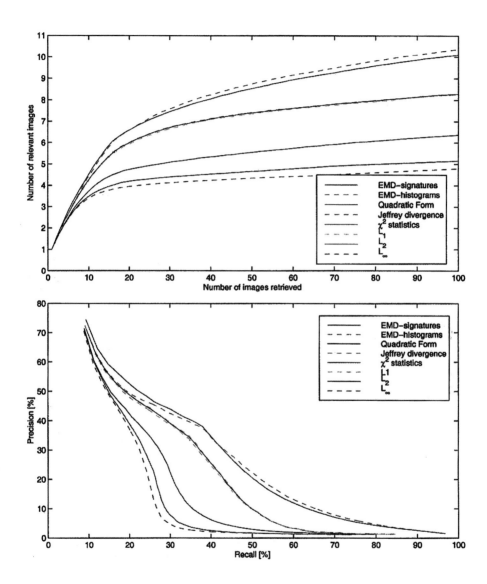

Figure B.10. Retrieving texture distributions using 24 filters and 512 bin histograms. *(This is a color figure)*

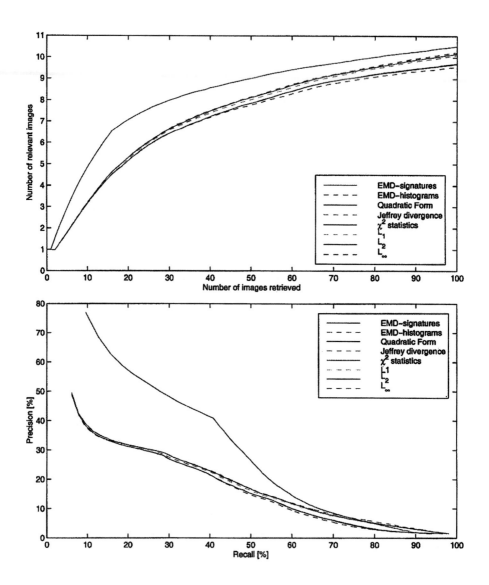

Figure B.11.   Retrieving texture distributions using 40 filters and 32 bin histograms. *(This is a color figure)*

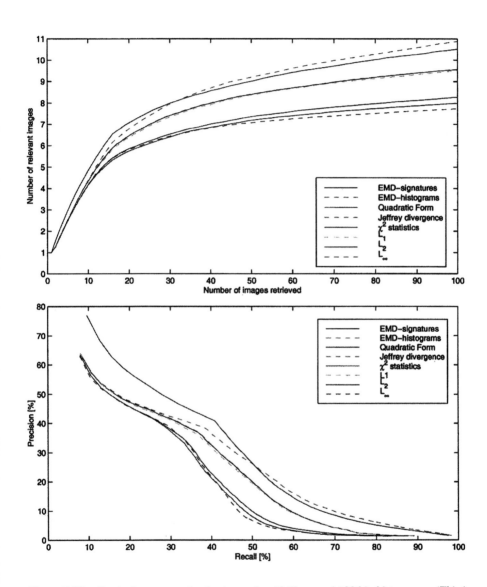

Figure B.12. Retrieving texture distributions using 40 filters and 128 bin histograms. *(This is a color figure)*

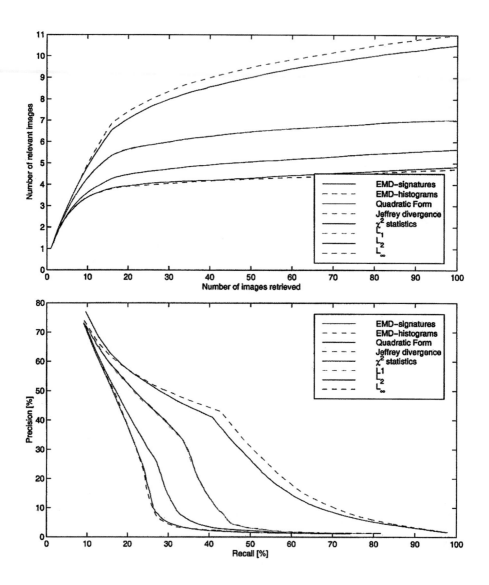

Figure B.13.   Retrieving texture distributions using 40 filters and 512 bin histograms. *(This is a color figure)*

References

[1] Edward H. Adelson and James R. Bergen. Spatiotemporal energy models for the perception of motion. *Journal of the Optical Society of America A - Optics and Image Science*, 2(2):284–299, February 1985.

[2] R. K. Ahuja, T. L. Magnanti, and J. B. Orlin. *Network Flows*. Prentice Hall, Englewood Cliffs, NJ, 1993.

[3] J. R. Bach, C. Fuller, A. Gupta, A. Hampapur, B. Horowitz, R. Humphrey, R. Jain, and C. Shu. Virage image search engine: an open framework for image management. In *SPIE Conference on Storage and Retrieval for Image and Video Databases IV*, volume 2670, pages 76–87, March 1996.

[4] S. Belongie, C. Carson, H. Greenspan, and J. Malik. Color- and texture-based image segmentation using EM and its application to content-based image retrieval. In *IEEE International Conference on Computer Vision*, pages 675–682, Bombay, India, January 1998.

[5] J. L. Bentley. Multidimensional binary search trees used for associative searching. *Communications of the ACM*, 18:509–517, September 1975.

[6] J. Bigün and J. M. du Buf. N-folded symmetries by complex moments in Gabor space and their application to unsupervised texture segmentation. *IEEE Transactions on Pattern Analysis and Machine Intelligence*, 16(1):80–87, 1994.

[7] A. C. Bovik, M. Clark, and W. S. Geisler. Multichannel texture analysis using localized spatial filters. *IEEE Transactions on Pattern Analysis and Machine Intelligence*, 12(12):55–73, 1990.

[8] T. Bozkaya and M. Ozsoyoglu. Distance-based indexing for high-dimensional metric spaces. *SIGMOD Record (ACM Special Interest Group on Management of Data)*, 26(2):357–368, May 1997.

[9] P. Brodatz. *Textures: A Photographic Album for Artists and Designers.* Dover, New York, NY, 1966.

[10] Kenneth L. Clarkson. Nearest neighbor queries in metric spaces. In *ACM Symposium on the Theory of Computing*, pages 609–617, May 1997.

[11] S. D. Cohen and L. J. Guibas. The earth mover's distance: Lower bounds and invariance under translation. Technical Report STAN-CS-TR-97-1597, Stanford University, November 1997.

[12] Scott Cohen. *Finding Color and Shape Patterns in Images.* PhD thesis, Stanford University, 1999. To be published.

[13] Thomas M. Cover and Joy A. Thomas. *Elements of Information Theory.* Wiley Series in Telecommunications. John Wiley & Sons, New York, NY, USA, 1991.

[14] Trevor F. Cox and Michael A. A. Cox. *Multidimensional Scaling.* Monographs on statistics and applied probability; 59. Chapman and Hall, London, 1994.

[15] A. Cumani. Edge detection in multispectral images. *Computer, Vision, Graphics, and Image Processing*, 53(1):40–51, 1991.

[16] Ralph B. D'Agostino and Michael A. Stephens, editors. *Goodness-of-fit techniques.* Statistics: Textbooks and Monographs, 68. Marcel Dekke, New York, 1986.

[17] G. B. Dantzig. *Linear Programming and Extensions.* Princeton University Press, Princeton, NJ, 1963.

[18] M. Das, E. M. Riseman, and B. A. Draper. FOCUS : Searching for multi-colored objects in a diverse image database. In *Proceedings of the IEEE Conference on Computer Vision and Pattern Recognition*, pages 756–761, June 1997.

[19] I. Daubechies. *Ten Lectures on Wavelets.* Number 61 in CBMS-NSF Regional Conference Series in Applied Mathematics. Society for Industrial and Applied Mathematics, Philadelphia, 1992.

[20] J. D. Daugman. Complete discrete 2-d Gabor transforms by neural networks for image analysis and compression. *IEEE Transactions on Acoustics, Speech, and Signal Processing*, 36:1169–1179, 1988.

[21] Silvano Di Zenzo. A note on the gradient of a multi-image. *Computer, Vision, Graphics, and Image Processing*, 33:116–125, 1986.

[22] Richard O. Duda and Peter E. Hart. *Pattern classification and scene analysis*. Wiley, New York, 1973.

[23] Ogle V. E. and Stonebraker M. Chabot: Retrieval from a relational database of images. *Computer*, pages 40–48, September 1995.

[24] C. Faloutsos, R. Barber, M. Flickner, J. Hafner, W. Niblack, D. Petkovic, and W. Equitz. Efficient and effective querying by image content. *Journal of Intelligent Information Systems*, 3:231–262, 1994.

[25] F. Farrokhnia and A. K. Jain. A multi-channel filtering approach to texture segmentation. In *Proceedings of the IEEE Conference on Computer Vision and Pattern Recognition*, pages 364–370, June 1991.

[26] Judy Feder. Image recognition and content-based retrieval for the world wide web. *Advanced Imaging*, 11(1):26–28, January 1996.

[27] David J. Field. Relations between the statistics of natural images and the response properties of cortical cells. *Journal of the Optical Society of America A*, 4(12):2379–2394, December 1987.

[28] Myron Flickner, Harpreet S. Sawhney, Jonathan Ashley, Qian Huang, Byron Dom, Monika Gorkani, Jim Hafner, Denis Lee, Dragutin Petkovic, David Steele, and Peter Yanker. Query by image and video content: The QBIC system. *IEEE Computer*, 28(9):23–32, September 1995.

[29] D. Forsyth, J. Malik, M. Fleck, H. Greenspan, and T. Leung. Finding pictures of objects in large collections of images. In *International Workshop on Object Recognition for Computer Vision*, Cambridge, UK, April 1996.

[30] D. Gabor. Theory of communication. *The Journal of the Institute of Electrical Engineers, Part III*, 93(21):429–457, January 1946.

[31] James E. Gary and Rajiv Mehrotra. Similar shape retrieval using a structural feature index. *Information Systems*, 18(7):525–537, October 1930.

[32] D. Geman, S. Geman, C. Graffigne, and P. Dong. Boundary detection by constrained optimization. *IEEE Transactions on Pattern Analysis and Machine Intelligence*, 12(7):609–628, 1990.

[33] A. Gersho and R. M. Gray. *Vector quantization and signal compression*. Kluwer Academic Publishers, Boston, MA, 1992.

[34] Gene H. Golub and Charles F. Van Loan. *Matrix Computations*. The Johns Hopkins University Press, Baltimore, Md, 1996. 3rd edition.

[35] J. Hafner, H. S. Sawhney, W. Equitz, M. Flickner, and W. Niblack. Efficient color histogram indexing for quadratic form distance functions. *IEEE Transactions on Pattern Analysis and Machine Intelligence*, 17(7):729–735, 1995.

[36] D. J. Heeger and J. R. Bergen. Pyramid-based texture analysis/synthesis. In *Computer Graphics*, pages 229–238. ACM SIGGRAPH, 1995.

[37] F. S. Hillier and G. J. Lieberman. *Introduction to Mathematical Programming*. McGraw-Hill, 1990.

[38] F. L. Hitchcock. The distribution of a product from several sources to numerous localities. *J. Math. Phys.*, 20:224–230, 1941.

[39] Thomas Hofmann, Jan Puzicha, and Joachim Buhmann. Unsupervised texture segmentation in a deterministic annealing framework. *IEEE Transactions on Pattern Analysis and Machine Intelligence*, 20(8):803–818, 1998.

[40] T. Y. Hou, A. Hsu, P. Liu, and M. Y. Chiu. A content-based indexing technique using relative geometry features. In *SPIE Conference on Image Storage and Retrieval Systems*, volume 1662, pages 59–68, 1992.

[41] F. Idris and S. Panchanathan. Review of image and video indexing techniques. *Journal of Visual Communication and Image Representation*, 8(2):146–166, June 1997.

[42] H. V. Jagadish. A retrieval technique for similar shapes. In *Proceedings of the 1991 ACM SIGMOD International Conference on Management of Data*, pages 208–217, Denver, Colorado, 1991. ACM Press.

[43] B. Jähne. *Digital Image Processing*. Springer, 1995.

[44] A. K. Jain and A. Vailaya. Image retrieval using color and shape. *Pattern Recognition*, 29(8):1233–1244, 1996.

[45] N. Karmarkar. A new polynomial-time algorithm for linear programming. In *Proceedings of the Sixteenth Annual ACM Symposium on Theory of Computing*, pages 302–311, Washington, D.C., April 1984.

[46] V. Klee and G. Minty. How good is the simplex algorithm. In O. Shisha, editor, *Inequalities*, volume III, pages 159–175. Academic Press, New York, N.Y., 1972.

[47] H. Knutsson and G. H. Granlund. Texture analysis using two-dimensional quadrature filters. *IEEE Computer Society Workshop on Computer Architecture for Pattern Analysis and Image Database Management*, 1983.

[48] E. Kreyszing. *Differential Geometry*. University of Toronto Press, Toronto, 1959.

[49] J. B. Kruskal. Multi-dimensional scaling by optimizing goodness-of-fit to a nonmetric hypothesis. *Psychometrika*, 29:1–27, 1964.

[50] S. Kullback. *Information Theory and Statistics*. Dover, New York, NY, 1968.

[51] Yingmei Lavin, Rajesh Batra, and Lambertus Hesselink. Feature comparisons of vector fields using earth mover's distance. In *IEEE/ACM Visualization '98*, pages 413–415, October 1998.

[52] T. S. Lee. Image representation using 2d Gabor wavelets. *IEEE Transactions on Pattern Analysis and Machine Intelligence*, 18(10):959–971, 1996.

[53] F. Liu and R. W. Picard. Periodicity, directionality, and randomness: Wold features for image modeling and retrieval. *IEEE Transactions on Pattern Analysis and Machine Intelligence*, 18(7):722–733, 1996.

[54] W. Y. Ma and B. S. Manjunath. Texture features and learning similarity. In *IEEE Conference on Computer Vision and Pattern Recognition*, pages 425–430, San Francisco, CA, 1996.

[55] J. Malik and P. Perona. Preattentive texture discrimination with early vision mechanisms. *Journal of the Optical Society of America A*, 7(5):923–932, May 1990.

[56] B. S. Manjunath and W. Y. Ma. Texture features for browsing and retrieval of image data. *IEEE Transactions on Pattern Analysis and Machine Intelligence*, 18(8):837–842, 1996.

[57] J. Mao and A.K. Jain. Texture classification and segmentation using multiresolution simultaneous autoregressive models. *Pattern Recognition*, 25:173–188, 1992.

[58] J. Marks, B. Andalman, P. Beardsley, B. Freeman, S. Gibson, J. Hodgins, T. Kang, B. Mirtich, H. Pfister, W. Ruml, J. Seims, and S. Shieberr. Design galleries: A general approach to setting parameters for computer graphics and animation. In *SIGGRAPH 97*, pages 389–400, Los Angeles CA, 1997.

[59] T.P. Minka and R.W. Picard. Interactive learning with a society of models. *Pattern Recognition*, 30(4):565–581, April 1997.

[60] G. Monge. Mémoire sur la théorie des déblais et des remblais. *Histoire de l'Académie des sciences de Paris, avec les Mémoires de mathématique et de physique pour la même année*, pages 666–704, 1781.

[61] N. M. Nasrabad and R. A. King. Image coding using vector quantization: A review. *IEEE Transactions on Communication*, 36(8):957–971, August 1988.

[62] W. Niblack, R. Barber, W. Equitz, M. D. Flickner, E. H. Glasman, D. Petkovic, P. Yanker, C. Faloutsos, G. Taubin, and Y. Heights. Querying images by content, using color, texture, and shape. In *SPIE Conference on Storage and Retrieval for Image and Video Databases*, volume 1908, pages 173–187, April 1993.

[63] Timo Ojala, Matti Pietikäinen, and David Harwood. A comparative study of texture measures with classification based on feature distributions. *Pattern Recognition*, 29(1):51–59, 1996.

[64] Greg Pass and Ramin Zabih. Histogram refinement for content-based image retrieval. In *IEEE Workshop on Applications of Computer Vision*, Sarasota, FL, 1996.

[65] S. Peleg, M. Werman, and H. Rom. A unified approach to the change of resolution: Space and gray-level. *IEEE Transactions on Pattern Analysis and Machine Intelligence*, 11:739–742, 1989.

[66] A. Pentland, R. W. Picard, and S. Sclaroff. Photobook: content-based manipulation of image databases. *International Journal of Computer Vision*, 18(3):233–254, June 1996.

[67] P. Perona. Deformable kernels for early vision. In *Proceedings of the IEEE Conference on Computer Vision and Pattern Recognition*, pages 222–227, June 1991.

[68] P. Perona and J. Malik. Scale-space and edge detection using anisotropic diffusion. *IEEE Transactions on Pattern Analysis and Machine Intelligence*, 12(7):629–939, 1990.

[69] R. W. Picard and T. P. Minka. Vision texture for annotation. *Multimedia Systems*, 3:3–14, 1995.

[70] C. Poynton. *A Technical Introduction to Digital Video*. John Wiley and Sons, New York, NY, 1996.

[71] Jan Puzicha, TThomas Hofmann, and Joachim Buhmann. Non-parametric similarity measures for unsupervised texture segmentation and image retrieval. In *Proceedings of the IEEE Conference on Computer Vision and Pattern Recognition*, pages 267–272, June 1997.

[72] S. T. Rachev. The Monge-Kantorovich mass transference problem and its stochastic applications. *Theory of Probability and its Applications*, XXIX(4):647–676, 1984.

[73] K. Rodden. How do people organise their photographs? In *BCS IRSG 21st Annual Colloquium on Information Retrieval Research*, GLasgow, England, 1999. to appear.

[74] Yossi Rubner, Leonidas J. Guibas, and Carlo Tomasi. The earth mover's distance, multidimensional scaling, and color-based image retrieval. In *Proceedings of the ARPA Image Understanding Workshop*, pages 661–668, New Orleans, LA, 1997.

[75] Yossi Rubner and Carlo Tomasi. Coalescing texture descriptors. In *Proceedings of the ARPA Image Understanding Workshop*, pages 927–935, Palm Springs, CA, February 1996.

[76] Yossi Rubner and Carlo Tomasi. Comparing the emd to other dissimilarity measures for color images. In *Proceedings of the DARPA Image Understanding Workshop*, pages 331–339, Monterey, CA, November 1998.

[77] Yossi Rubner and Carlo Tomasi. Texture metrics. In *Proceedings of the IEEE International Conference on Systems, Man and Cybernetics*, pages 4601–4607, San Diego, CA, October 1998. IEEE Systems, Man and Cybernetics Society.

[78] Yossi Rubner, Carlo Tomasi, and Leonidas J. Guibas. Adaptive color-image embeddings for database navigation. In *Asian Conference on Computer Vision*, pages 104–111, Hong-Kong, China, January 1998.

[79] Yossi Rubner, Carlo Tomasi, and Leonidas. J. Guibas. The earth mover's distance as a metric for image retrieval. Technical Report STAN-CS-TN-98-86, Computer Science Department, Stanford University, September 1998.

[80] Yossi Rubner, Carlo Tomasi, and Leonidas J. Guibas. A metric for distributions with applications to image databases. In *IEEE International Conference on Computer Vision*, pages 59–66, Bombay, India, January 1998.

[81] Y. Rui, T. Huang, and S. Mehrotra. Content-based image retrieval with relevance feedback in mars. In *IEEE International Conference on Image Processing*, pages II:815–xx, 1997.

[82] E. J. Russell. Extension of Dantzig's algorithm to finding an initial near-optimal basis for the transportation problem. *Operations Research*, 17:187–191, 1969.

[83] M. Ruzon and C. Tomasi. Color edge detection with the compass operator. In *Proceedings of the IEEE Conference on Computer Vision and Pattern Recognition*, 1999. to appear.

[84] P. Saint-Marc, J. S. Chen, and G. Medioni. Adaptive smoothing: A general tool for early vision. *IEEE Transactions on Pattern Analysis and Machine Intelligence*, 13(6):514–529, 1991.

[85] S. Santini and R. Jain. Similarity queries in image databases. In *Proceedings of the IEEE Conference on Computer Vision and Pattern Recognition*, pages 646–651, San Francisco, CA, 1996.

[86] G. Sapiro and D.L. Ringach. Anisotropic diffusion of multivalued images with applications to color filtering. *IEEE Transactions on Image Processing*, 5:1582–1586, 1996.

[87] H. C. Shen and A. K. C. Wong. Generalized texture representation and metric. *Computer, Vision, Graphics, and Image Processing*, 23:187–206, 1983.

[88] Roger. N. Shepard. The analysis of proximities: Multidimensional scaling with an unknown distance function, i and ii. *Psychometrika*, 27:125–140,219–246, 1962.

[89] Roger. N. Shepard. Toward a universal law of generalization for psychological science. *Science*, 237:1317–1323, 1987.

[90] R. Sibson. Studies in the robutness of multidimensional scaling: Procrustes statistics. *Journal of the Royal Statistic Society*, 40:234–238, 1978.

[91] D. Slater and G. Healey. The illumination-invariant recognition of 3D objects using local color invariants. *IEEE Transactions on Pattern Analysis and Machine Intelligence*, 18(2):206–210, 1996.

[92] J. R. Smith. *Integrated Spatial and Feature Image Systems: Retrieval, Analysis and Compression*. PhD thesis, Columbia University, 1997.

[93] N. Sochen, R. Kimmel, and R. Malladi. A general framework for low level vision. *IEEE Transactions on Image Processing*, 7(2):310–317, 1998.

[94] J. Stolfi, 1994. Personal communication.

[95] M. Stricker and A. Dimai. Color indexing with weak spatial constraints. In *SPIE Conference on Storage and Retrieval for Image and Video Databases IV*, volume 2670, pages 29–40, March 1996.

[96] M. Stricker and M. Orengo. Similarity of color images. In *SPIE Conference on Storage and Retrieval for Image and Video Databases III*, volume 2420, pages 381–392, February 1995.

[97] Michael J. Swain and Dana H. Ballard. Color indexing. *International Journal of Computer Vision*, 7(1):11–32, 1991.

[98] Y. Takane, F. W. Young, and J. Leeuw. Nonmetric individual differences multidimensional scaling: an alternating least squares method with optimal scaling features. *Psychometrika*, 42:7–67, 1977.

[99] H. Tamura, T. Mori, and T. Yamawaki. Textural features corresponding to visual perception. *IEEE Transactions on Systems, Man and Cybernetics*, 8:460–473, June 1978.

[100] W. S. Torgerson. *Theory and Methods of Scaling*. John Wiley and Sons, New York, NY, 1958.

[101] H. Voorhees and T. Poggio. Detecting textons and texture boundaries in natural images. In *IEEE International Conference on Computer Vision*, pages 250–258, 1987.

[102] Harry Voorhees and Tomaso Poggio. Computing texture boundaries from images. *Nature*, 333:364–367, 1988.

[103] Brian A. Wandell. *Foundations of Vision*. Sinauer Associates, Sunderland, MA, 1995.

[104] Andrew B. Watson. The cortex transform: rapid computation of simulated neural images. *Computer, Vision, Graphics, and Image Processing*, 39:311–327, 1987.

[105] M. Werman, S. Peleg, and A. Rosenfeld. A distance metric for multidimensional histograms. *Computer, Vision, Graphics, and Image Processing*, 32:328–336, 1985.

[106] G. Wyszecki and W. S. Stiles. *Color Science: Concepts and Methods, Quantitative Data and Formulae.* John Wiley and Sons, New York, NY, 1982.

[107] Xuemei Zhang and Brian A. Wandell. A spatial extension of CIELAB for digital color image reproduction. In *Proceedings of SID*, pages 731–734, San Diego, CA, May 1996.

[108] K. Zikan. *The Theory and Applications of Algebraic Metric Spaces.* PhD thesis, Stanford University, 1990.

Index

CPSIA information can be obtained at www.ICGtesting.com
Printed in the USA
LVOW070627040412

276076LV00004B/37/A